Fostering Women's Engagement in STEM Through Education

A Cross-Cultural Academic-Industry Journey

Edited by
Esyin Chew and Anwar P.P. Abdul Majeed

CRC Press
Taylor & Francis Group
Boca Raton London New York

CRC Press is an imprint of the
Taylor & Francis Group, an **informa** business

First edition published 2025
by CRC Press
2385 NW Executive Center Drive, Suite 320, Boca Raton FL 33431

and by CRC Press
4 Park Square, Milton Park, Abingdon, Oxon, OX14 4RN

CRC Press is an imprint of Taylor & Francis Group, LLC

ISBN: 978-1-032-84717-7 (hbk)
ISBN: 978-1-032-84718-4 (pbk)
ISBN: 978-1-003-51462-6 (ebk)

DOI: 10.1201/9781003514626

Typeset in Times
by KnowledgeWorks Global Ltd.

Contents

Preface   vii
Editor Bios   ix
List of Contributors   xi

1 **Introduction to Women in STEM**   1
   Esyin Chew

2 **EUREKA STEM Robotics and Artificial Intelligence
   Initiatives in Wales**   16
   Chow Siing Sia and Esyin Chew

3 **Industrial Sponsored STEM Star @UiTM**   31
   Sukreen Hana Herman and Rozita Jailani

4 **Exploring Challenges, Best Practices, and the Road
   Ahead for Digital Making Skillsets in STEM Education**   46
   Nurul Hazlina Noordin

5 **Congkak Quest: Enhancing STEM Skills through
   Game-Based Learning and Cultural Exchange**   70
   *Zati Hakim Azizul Hasan, Mas Sahidayana Mohktar,
   and Siti Nursheena Mohd Zain*

6 **Implication, Challenges, and Moving Forward**   86
   Esyin Chew and Anwar P.P. Abdul Majeed

Index   99

Preface

Promoting gender equality and diversity in the science, technology, engineering, and mathematics (STEM) fields has become a global imperative. Despite progress over the past century, women remain significantly underrepresented in STEM education and careers worldwide. This gender gap represents a loss of critical talent, innovation, and socioeconomic development potential.

This book, *Fostering Women's Engagement with STEM through Education: A Cross-Cultural Academic-Industry Journey*, brings together insights and experiences from a pioneering programme, Partnership for Innovation in Employability (PIE), supported by the British Council. Spearheaded by the EUREKA Robotics Centre at Cardiff Metropolitan University, along with Universiti Malaya, Universiti Teknologi MARA, Universiti Malaysia Pahang Al-Sultan Abdullah, and BINUS University, the PIE programme aims to build strategic academia-industry links across different countries to empower women in STEM through skills development, mentoring, and collaborative opportunities.

The book traces the PIE programme's cross-cultural journey, providing a thought-provoking account of the status, challenges, and initiatives related to women in STEM across various regions. It delves into specific experiences and case studies from the programme in Wales, Malaysia, and beyond, covering topics ranging from STEM robotics' public impact to game-based learning.

The authors, representing a powerful collective of women and men in STEM, offer candid reflections on the implications, challenges, and future vision emerging from the PIE programme. Their insights underscore the crucial role of fostering women's engagement with STEM through education and cross-sectoral partnerships.

This book serves as a clarion call for accelerating concerted efforts to bridge the STEM gender divide. It provides policymakers, institutions, educators, and advocates with valuable lessons and inspiration to forge ahead in unlocking the full potential of women in STEM. Ultimately, advancing gender equality in STEM is not just a matter of equity; it is essential for driving research, innovation, and socioeconomic progress for the benefit of all.

We hope this book enlightens and galvanises readers to take forward the movement for women's empowerment and gender equality in STEM, from classrooms to laboratories, from academia to industry, and from local to global arenas. The journey is far from over, but through sharing knowledge and uniting efforts, a gender-equal future in STEM is within reach.

Esyin Chew and Anwar P.P. Abdul Majeed (Eds)

Editor Bios

Dr Esyin Chew is the Director of the EUREKA Robotics Centre, which has more than 120 robots and is one of the 11 specialist robotics centres in the UK for research facilities. Esyin has successfully developed eight Wales Tokku Zones in the UK and Asia with real-world social-educational and healthcare robotics experiments. She is a Reader in Robotics and Educational Technologies providing consultancy and training; a keynote speaker for many international conferences in robotics and EduTech; and an editor and reviewer for peer-reviewed journals and conference proceedings in the related field. Esyin is a creative academic investigating humanoid service robotics with machine learning capability for impact case studies and has led government- and industry-funded projects from the EU, Australia, Malaysia, and the UK, as well as over 150 R&D outputs and six national policy impact publications.

Anwar P.P. Abdul Majeed is an Associate Professor at the School of Engineering and Technology, Sunway University, with experience in transnational education. He is a Chartered Engineer with the Institute of Mechanical Engineers (IMechE), a member of the Institution of Engineering and Technology (IET), and a Senior Member of Institute of Electrical and Electronics Engineers (IEEE). With over 200 publications and editorial roles in prestigious journals and book series, Dr Anwar is committed to advancing knowledge in the application of machine learning and deep learning in different domains. He is also actively engaged in professional societies and alumni networks, and contributes to the scientific community through his involvement in the Young Scientists Network of the Academy of Sciences Malaysia.

Contributors

Anwar P.P. Abdul Majeed
School of Engineering and
    Technology
Sunway University
Petaling Jaya, Malaysia

Zati Hakim Azizul Hasan
Faculty of Computer Science
Universiti Malaya
Kuala Lumpur, Malaysia

Esyin Chew
Cardiff School of Technologies
Cardiff Metropolitan University
    Cardiff, UK

Sukreen Hana Herman
College of Engineering
Universiti Teknologi MARA
Shah Alam, Malaysia

Rozita Jailani
College of Engineering
Universiti Teknologi MARA
Shah Alam, Malaysia

Siti Nursheena Mohd Zain
Faculty of Science
Universiti Malaya
Kuala Lumpur, Malaysia

Mas Sahidayana Mohktar
Faculty of Engineering
Universiti Malaya
Kuala Lumpur, Malaysia

Nurul Hazlina Noordin
Faculty of Electrical Electronics
    Engineering Technology
Universiti Malaysia Pahang
    Al-Sultan Abdullah
Pekan, Malaysia

Chow Siing Sia
Cardiff School of Technologies
Cardiff Metropolitan University
Cardiff, UK

Introduction to Women in STEM

1

Esyin Chew

1.1 OVERVIEW OF THE IMPORTANCE FOR GENDER EQUALITIES AND DIVERSITY IN STEM

Reading the title of this article is striking: 'Being in science and at the same time being a woman is difficult*': Academic women's experiences of gender inequalities in STEM* (Sidelil et al., 2023). I believe this is the heart-felt realism of all authors of this book and, possibly, you, who are considering and reading our experiences. STEM is the universal terminology representing science, technology, engineering, and mathematics. Progressive scientists and policymakers have identified a common problem all over the globe: *there is a strong gender gap in the world of STEM* (STEM Women, UN, 2023). Challenging women's engagement with STEM in higher education reveals its gap and the importance of bridging that gap, both in fundamental culture and good practices. Through cross-cultural education with sociological imagination and history (Wright, 1959), gender equality and diversity are evolved in the past and present. For instance, the first scientific woman in mankind history to receive a PhD was Elena Lucrezia Cornaro Piscopia in 1678, who went against the Catholic Church dogma, which holds that women are not fit to handle complicated logic, by becoming a mathematics and physics scholar

DOI: 10.1201/9781003514626-1

(Elkington, 2022; Guernsey, 1999). The present 2024 QS top five universities in the world (QS, 2024) –Massachusetts Institute of Technology (MIT), Cambridge, Oxford, Harvard and Standford – have profoundly accelerated the making of women's history in higher education, though the progress was complex, tardy but impactful.

Over a hundred and fifty years ago, the first woman to graduate from MIT was Ellen Swallow with a Bachelor of Science in 1873 (MIT, 2017). In 1896, Jane Stanford, the University of Stanford's founder and administrator, paved the way for a cohort of pioneering women in an academic world dominated by men (Stanford, 2016). In 1920, Annie Rogers and Ivy Williams were remarkably the first women awarded Oxford University degrees (Oxford, 2020). Between 1923 and 1924, in the wake of World War I, almost 400,000 women from Wales signed a petition appealing to the women of America, calling for world peace. Subsequently, the Welsh League of Nations Union (WLNU) was established, led by Annie Hughes-Griffiths, the WLNU Chair and leader of the Women Peace Petition (Rhyfeddol & Menywod, 2023; WCIA, 2023). The Welsh women's two-month 'Peace Tour' of the United States built support through American women's organisations involving over 60 million people. The aim of the WLNU was 'to promote international cooperation and to achieve international peace and health' and, subsequently, the Welsh women are believed to have inspired the founding of the League of Nations, the predecessor of the United Nations with a list of 52 members states, to bring an end to World War II (UN, 2024).

Around the same period during World War II, Harvard University allowed women students into their classrooms for the first time and, in 1948, Helen Maud Cam became the first female faculty member there. Astronomy Professor Cecilia Payne-Gaposchkin became the first woman member of the Faculty of Arts and Sciences, and later became the first female to head a department at Harvard (2024). In the same year, the University of Cambridge was the last British university to promote gender equality in higher education by awarding its first female recipient of a Cambridge degree (honorary) to the then Queen Mother, Elizabeth Bowes-Lyon (Cambridge, 2024).

Despite the long historical events in the world-leading universities that helped shape today's higher education, much work still needs to be done, in particular with regard to gender inequality and diversity in STEM fields at all levels (García-Holgado et al., 2019). The World Economic Forum (2020, 2023) summarised the trending issues as: (1) women employees and students are considerably underrepresented in STEM-related disciplines worldwide; (2) on average, approximately 30% of the world's researchers are women;

(3) less than one-third of women students choose to pursue STEM subjects in higher education, including mathematics and engineering; (3) women in STEM disciplines generally publish less and always receive less pay in the global perspectives. From the global horizon, women have been underrepresented in many STEM fields in most countries outside the US and UK, including:

a. **STEM women in Latin America**: the lack of women in STEM is a severe problem in Latin America due to the cultural norms and biases. The Faculty of Mathematical and Physical Sciences at the University of Chile has created the Gender Equality Admissions Program to admit an extra 40 women into STEM. The W-STEM project seeks to improve strategies and mechanisms for attracting, accessing, and guiding women in Latin America into STEM higher education programmes. A model for bridging the gender gap in STEM was proposed along with mentoring female students in engineering as a way of caring (Bastarrica et al., 2018; García-Holgado et al., 2019, 2022).

b. **STEM women in Africa**: a critical interpretive analysis of the South African situation for women in STEM suggested the preliminary findings provide valuable insights into the barriers, opportunities, and strategies for promoting the adoption of immersive technology for STEM education in Nigeria and potentially other African countries (Pretorius et al., 2010; Sakpere et al., 2024). In spite of the considerable advancements for women in Africa that are 'reform' but not 'transformation,' the gender gap in STEM is still a severe reality. Women's priceless perspectives and contributions to scientific discoveries are directly hampered by the gender social-cultural biases in the region (Founou et al., 2023; Mkhize, 2023).

c. **STEM women in Australia**: researchers propounded the notion of a 'women crisis' to challenge the biases because Australia-based academics are underrepresented in major STEM conferences, resulting in potential ripple effects on the careers of female scholars for employability, both locally or abroad (Capel et al., 2016; McKay & Buchanan, 2021). Despite the Decadal Plan for Women in STEM of the Australian Academy of Science providing an overview of 337 programmes and initiatives offered nationally by joint efforts from academia, industry, and government, the absence of evidence of the effectiveness of Australian gender equity in STEM initiatives is suggested (McKinnon, 2022).

d. **STEM women in the Middle East**: due to the stereotyping of Middle Eastern women, professional Arab women in STEM, especially young women in engineering and computer science, continue to experience citizenship quandaries, family and community responsibilities, and gender-racial prejudice in the pre- and post-pandemic studies (Alzaabi et al., 2021; Koblitz, 2016; Patterson et al., 2021; Pilotti et al., 2024). Studies including gendered micro-aggressions at work and how women in STEM navigated them with self-efficacy in the Middle East region are worth referencing (Makarem & Metcalfe, 2023).

e. **STEM women in Asia**: both Singapore and Malaysia commenced the STEM for Women initiatives early in 2018 to set up the National STEM Association (by a woman who later become the Minister of Higher Education) and promote gender discourse of women's industrial leadership in STEM, which has been hampered due to Southeast Asian's patriarchal gender roles and culturally constituted organisational perceptions of women and their leadership competence further hindering the potential for women leaders in STEM careers (Dutta, 2018; NSA, 2018). Joint efforts from (1) raising cross-countries public awareness about the important role of STEM women; (2) the quantity and quality of STEM women graduates or workforce; (3) international mobility, ways to attract international students, faculty and researchers, and (4) engaging strategies with STEM experts from Bangladesh, India, Indonesia, Malaysia, Nepal, Pakistan, Singapore, Sri Lanka, and Thailand are linked to fast-paced economic development in Asia (Chowdhury et al., 2022; Varma et al., 2023; Yamada, 2023;)

STEM issues in the UK and US are prevalent and initiatives are in the early days. The UK started a national STEM Learning Centre in 2004, 14 years before Malaysia (STEM, 2024). Based on the UK's recent Higher Education Statistics Agency (HESA) data, only 31% of STEM students in UK universities are women – see the subject breakdown at Figure 1.1 (STEM Women, 2023).

The House of Lords' Select Committee on Artificial Intelligence (AI) (2018) stated the lack of gender diversity in AI development led to a significant impact on the way that AI systems are designed and developed. To address the gender disparity between boys and girls studying STEM in the UK, measurements and investment include the funding of a £27 million provision for mathematics in 3,000 schools, and £40 million grants to establish Further Education Centres of Excellence across

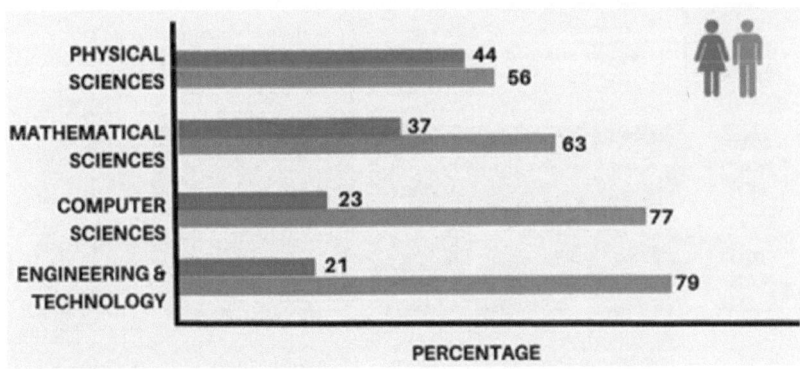

FIGURE 1.1 Core STEM subject statistics (STEM Women, 2023).

the UK to train mathematics educators and share the national best practice (UK Parliament, 2018). In parallel, I would suggest an enhanced STEM-robotics educational curriculum from pre-school up to higher education for developing graduates, especially for girls and women, with the skills that will never be replaced by AI in the fourth industrialisation (Chew-UK Parliament, 2018).

In terms of employability and industrial sector, the Alan Turing Institute criticises how female-founded AI startups in the UK received six times less funding than male founders between 2012 and 2022 (Huge, 2023). Bridging the digital gender divide in the exponential growth of AI and data has become paramount, and includes bias in robotics products and algorithmic systems. Research has shown that women comprise a smaller portion of the workforce, and girls and women face barriers to pursuing these fields of study due to a lack of training, and students disproportionately favoured non-STEM programmes over STEM programmes prior to the pandemic (Hutchinson, 2014; Pilotti et al., 2024; STEM Women, 2024). 'Achieving gender equality will significantly advance the STEM labour force, research and innovation, enhance the economy, and reduce the risk of women's social exclusion to the benefit of society.' (Fatourou et al., 2019, p. 1)

Employment in STEM fields remains disproportionately dominated by men, especially in engineering and maths. Due to the pandemic, women's careers were interrupted and nearly 3 million women have left the US job market with limited opportunities to return to the workforce (Quinn & Hayashi, 2021). According to a poll conducted by the US Census Bureau, only 27% of STEM jobs were held by women whereas only 29% of the current STEM workforce are women and further investment from the government is required including the pilot by the UK Minister for Women

STRATEGIC OBJECTIVES	Reduce inequalities and promote learning and creative societies in the digital age through quality education for all	Work towards sustainable societies by preserving the environment through the promotion of science, technology and the natural heritage
GENDER EQUALITY FOCUS	ACHIEVE GENDER EQUALITY IN AND THROUGH EDUCATION	EMPOWER WOMEN IN SCIENCE AND TECHNOLOGY FOR ENVIRONMENTAL ACTION
AREAS OF WORK	ACCESS — CURRICULUM & PEDAGOGIES — STEM	SCIENTIFIC LEADERSHIP — ECOSYSTEMS & BIODIVERSITY — WATER & OCEANS

Areas of work details:

- Access to equitable and gender-responsive education and learning, including in science, technology, engineering and mathematics (STEM) is widened
- Laws, policies & strategies for gender equality in and through education developed and implemented
- Gender-responsive teaching, content and learning opportunities are increased
- Gender gap in learning achievements and completion rates reduced

▶ *Contributing to SDG Targets 4.1, 4.3, 4.5, 4.7*

- Women's careers in decision-making positions advanced and role models promoted
- Gender transformative science, technology and innovation (STI) policies and Open Science policies enhanced
- Women's and girls' resilience and action for sustainable management of natural resources reinforced
- Gender responsive water and ocean's management and governance promoted

▶ *Contributing to SDG Target 5.5*

CROSS CUTTING THEME	WOMEN PEACE AND SECURITY:

ACTIONS	• Awareness raised through advocacy campaigns and public debates • Capacities and skills strengthened • National laws, policies and strategies elaboration and revision supported

FIGURE 1.2   UNESCO Global Priority Gender Equality Framework Part I (UN, 2023).

and Equalities to get women back into STEM workforce (IET, 2024; UK Government, 2023; US Census, 2021). The percentage is even smaller for managerial and senior-level positions. Hence, the United Nations (2023) urged that intersecting global crises cannot be solved without gender equality and invested 16% of the United Nations Educational, Scientific and Cultural Organization's (UNESCO's) budget in gender equality initiatives, which is among the highest percentage across the United Nations' budgets to reduce inequalities through quality education, to empower women in STEM to bring about sustainable environmental action, to promote inclusion, combat gender-based violence, and bridge the digital divide, as depicted in Figures 1.2 and 1.3.

Build inclusive, just and peaceful societies by promoting freedom of expression, cultural diversity, education for global citizenship, and protecting the heritage

Foster a technological environment in the service of humankind through the development and dissemination of knowledge and skills and the development of ethical standards

PROMOTE INCLUSION AND COMBAT GENDER-BASED VIOLENCE

BRIDGE THE DIGITAL GENDER DIVIDE

DISCRIMINATION & STEREOTYPES	FREEDOM OF EXPRESSION & CREATION	DECENT WORK & SOCIAL PROTECTION	ETHICAL STANDARDS FOR AI & DIGITAL PLATFORMS	MEDIA & INFORMATION LITERACY	DIGITAL SKILLS & COMPETENCIES

- Gender-based threats, violence, discrimination and stereotypes tackled, including in the digital environment
- Learners and learning environments equipped to be safe and healthy
- Women's working conditions, employment opportunities and economic, cultural and social rights improved
- Gender equality promoted in and through the media and cultural sectors

▶ *Contributing to SDG Targets 4.2a, 5.1, 5.2, 5.6, 8.5, 8.8, 16.10*

- Ethical standards addressing gender bias in digital technologies and artificial intelligence (AI) systems developed and implemented
- Equitable and gender responsive access to and use of information and knowledge fostered in the digital environment
- Women and girls empowered with digital skills and competencies
- Women's leadership in AI and the digital environment supported

▶ *Contributing to SDG Target 5.b*

FIGURE 1.3 UNESCO Global Priority Gender Equality Framework Part II (UN, 2023).

Quoting the urging message from the Director-General of the United Nations, Audrey Azoulay (2023), who summarised the importance of gender equalities and diversity in STEM:

> *Nearly 300 years. It will take nearly 300 years to achieve gender equality worldwide, if we do not speed up our efforts. Indeed, the present situation remains worrying. Not only is progress too slow, where girls and women have been brutally deprived of their right to learn and teach, have led to major setbacks. Now is the time to send a simple message: we cannot accept half of humanity being left behind for another three centuries. This is why UNESCO has made gender equality a priority throughout its mandate, and why we are mobilizing our tools and our partners.*

1.2 INTRODUCTION TO THE PIE PROGRAMME AND ITS GOALS

The EUREKA Robotics Centre, Cardiff Metropolitan University (CMET), commenced the STEM education for underprivileged communities in 2018 through robotics innovation. Given the landscape of the gender equality and diversity agenda, we responded to UNESCO's Global Priority Gender Equality framework by introducing the PIE I (Partnership for Innovation in Employability) for Women in STEM programme in 2022, PIE II (Partnership for Innovation in Employability-Education) for Women in STEAM-H with the additional 'A' for agriculture and 'H' to represent women in healthcare in 2023. Starting out as a group of mothers and a fatherly thorn among the roses in the UK, Malaysia, and Indonesia, the PIE programme (PIE, 2024) is linked with the UK-Malaysia Universities Consortium that consists of 16 UK universities and 20 Malaysian universities. Together, we aim to amplify and expand its reach and multiply gender-transformative results for the benefit of women and girls, men and boys.

Inspired by the Welsh women's spirit of peace and health with education and mobility (WCIA, 2023), 'from home to home, hearth to hearth,' the Project Investigator took a human approach to influence and impact the PIE members and members' institutions with ethical robotics and AI integration, and local-global mentorship. Capitalising six public higher educational institutions (CMET, University of York, Universiti Teknologi MARA (UiTM), Universiti Malaya (UM), Universiti Malaysia Pahang Al-Sultan Abdullah (UMPSA) and BINUS University) and two industries (ALTY Hospital and Robopreneur Ltd) across the UK, Malaysia and Indonesia, the PIE I programme aims to enhance the employability of women in STEM through the following objectives:

1. Building capacity to establish strategic links across the six universities in the UK and Malaysia, as well as between academia and industry through the current universities' industrial partnership.
2. Encouraging development of strategically beneficial and internationally significant collaborative relationships and STEM projects, i.e. industrial service robotics and AI development for the healthcare sector in Wales, and Women for STEM workshops delivery for schools.
3. Providing funding for short-term strategic insight placement of women from partnering academic institutions in Malaysia and

Indonesia into external organisations, which can lead to productive, long-term relationships.

4. Contributing to graduate employability and MSc in Robotics and AI curriculum development by instilling the Cardiff Met EDGE – a series of ethical, digital, global, and entrepreneurial skills – to develop confidence, resilience, and experiences for women in STEM that will prepare PIE programme participants for life after university and through to employment.

The Jisc's Employability Toolkit[1] is adapted as the fundamental rationale of the PIE programme, including the following. (1) **Describing an Employable Learner [Obj 1]** with robotics and AI employability roadshows and workshops in higher educational institutions, primary schools, high schools, and in public places like museums and castles. Placement interview and recruitment for undergraduates, postgraduates, and early career researchers are conducted. (2) **Embedding Technology from Employability into Delivery [Obj 2 & 3]** with women academics in STEM placement in Cardiff and York. (3) **Widening Support for Employability [Obj 2 & 3]** with robotics and AI field works in the UK and Malaysia care homes and hospitals, for PIE members to be invited to join EUREKA Robotics Centre's large scale implementation of service robots in Wales and globally. (4) **Incorporating Employability into Curriculum Design [Obj 4]** with shared knowledge exchange for employability across the six aforementioned universities. Peer review for the newly offered MSc in Robotics and AI at Cardiff School of Technologies with possibilities of joint dissertation supervisions and dual MPhil/PhD programmes among the partners.

The PIE I programme, therefore, is not merely about getting jobs for women in STEM, it is about a broader set of skill and attribute developments that will enable both women graduates and academic staff to be successful throughout their working life. This cross-countries joint programme will demonstrate the impact of enhancing graduate employability through knowledge sharing, training, and innovation across the areas of STEM research, teaching and learning, and student experience through robotics and AI with the below justifications:

- A short-term measurable and tangible catalyst for women from four partner universities to have placements in the two UK universities – CMET and the University of York – as well as an international supervisory team for solving real-world problems with the current healthcare robotics projects at the EUREKA Robotics Centre and York Robotics Lab.

- A medium-term implication is the joint MSc in Robotics & AI dissertation supervision to enhance international employability.
- A long-term self-sustainable partnership that will contribute to the consortium and value creation to UK, Malaysia, Indonesia, Pakistan, China, and globally, in the context of women in STEM employability.

The PIE II programme aims to address the above gap by (1) developing interdisciplinary dual PhD and micro-credential courses in aquaculture and healthcare applications for B40 students and female staff who work in hospitals; (2) delivering interdisciplinary public impact seminars for female healthcare staff and girls in school; and (3) empowering women leadership in the workplace with global mobility. This approach offers a real opportunity to address the challenges in the healthcare sector and an underrepresented gender group (B40). It aims at engaging women and girls in multi- and interdisciplinary mobility and development opportunities, breaking down barriers that have traditionally existed between different subjects (such as nursing, elderly care and AI, aquaculture and robotics), and offering programme participants opportunities to make new cross-disciplinary connections.

Overall, the PIE programme adapts the rising tide of AI and robotics to nurture innovation and facilitate interaction between participating universities and their industry partners. It allow partners to establish links with one another which it will enable through funding for short-term placement opportunities, and projects deliveries aimed at developing new cross-disciplinary relationships (healthcare, computer science, engineering, ethics, and design), encouraging the development of strategically beneficial collaborative projects in a supportive environment for women in STEM towards global facing.

1.3 SETTING THE CONTEXT FOR THE SUBSEQUENT CHAPTERS

Fostering women's engagement with STEM through education, this book produces a challenging and thought-provoking landscape of women in STEM history, global practices, and lessons learnt from a cross-cultural academic-industry journey. This book is timely because, as the various authors concur, we are confronted already with the realism of 300 years of slow progress in gender equality worldwide, from which extraordinarily accurate, as urged by UNESCO (2023). The content of the six thought-provoking chapters is outlined as follows.

This chapter introduces the background and horizon of women in STEM, from the history of women in higher education to women around the world participating in and leading STEM challenges and initiatives over the past decades. It provides an overview of the PIE programme and its aims through the support of the British Council, which brought together a powerful team of PIE women in STEM from the UK and Malaysia. These women collaborate, with mutual respect and impact, to change the world around them through the PIE ethos.

Chapter 2 highlights the EUREKA Robotics Centre's STEM Robotics Public Impact in Wales. It describes the STEM robotics initiatives by CMET's EUREKA Robotics Centre on the public impact it had in Wales in the past seven years of women efforts for women. This immediately prompts ideas and practical actions to bring the PIE agenda forward in Wales and globally.

Chapter 3 suggests a self-sustainable industrial-sponsored STEM Star at UiTM. This chapter delivers the STEM initiatives by UiTM, sponsored by different organisations, to increase STEM awareness within the heart of Malaysia, impacting thousands of girls' lives. A EUREKA Nexus Lab was established and launched at UiTM for dual MPhil/PhD research collaboration for joint women supervision.

Chapter 4 creatively reports on the government-supported Ninja STEM intervention in the east coast of Malaysia by UMPSA. The chapter delves into how this intervention changed the experience for the east coast compared with west coast and central Malaysia for Technical and Vocational Education and Training (TVET) and underprivileged communities.

Chapter 5 details the Malaysia-Wales STEM experience, from traditional games like Congkak to robots, through the STEM Centre within the best university in the country. This unique experience used a Malaysian past time to facilitate the understanding of STEM among a group of school children in Wales. The play-based education programme, Let's go to Mummie's Lab (LGTML) is inspiring and can be the world's reference for the slogan: 'Heads on, Hands on and Hearts On,' meaning the children think about what they are doing, get involved, and are interested and enjoy participating in the STEM activities (LGTML, 2022).

Chapter 6 concludes with the implication, challenges, and way forward for the PIE programme. This chapter summarises the various experiences and shares the challenges as well as the means to drive STEM in both countries moving forward. The lessons learnt from the journey, as well as practical aspirations, are particularly impactful. The shape of the book is helpful, drawing on a number of experts who provide complementary insights from their contribution to women in STEM as specialists in their fields. It is impossible to account for all the effects of the PIE programme's intervention, particularly in a short discussion like this. What follows is more crucially a long-term, self-sustainable PIE programme to a wider participation for at least the next five years.

Although women in STEM are underrepresented, we do not take the approach of radical feminism to evoke identity threat and cause negative outcomes. Instead, we can all agree that humanity, over the gender disparity and equality, is rooted truly in PIE for women in STEM.

From measurements to actions in PIE, we return to UNESCO's urge and echo the passionate message from Audrey Azoulay, the Director-General of United Nations (2023):

> Through all these actions, this book tells us how far we have come; but, more fundamentally, it warns us of the many efforts we still have to make. Let's not wait 300 more years to react. (pp. 5)

NOTE

1   https://www.jisc.ac.uk/guides/employability-toolkit

REFERENCES

Alzaabi, I., Ramírez-García, A., & Moyano, M. (2021). Gendered stem: A systematic review and applied analysis of female participation in stem in the United Arab Emirates. *Education Sciences*, *11*(10), 573.

Bastarrica, M.C., Hitschfeld, N., Samary, M.M., & Simmonds, J. (2018). Affirmative action for attracting women to STEM in Chile. In *Proceedings of the 1st international workshop on gender equality in software engineering (GE '18)* (pp. 45–48). Association for Computing Machinery. https://doi-org.ezproxy.cardiffmet.ac.uk/10.1145/3195570.3195576

Cambridge. (2024). *The rising tide: Women at Cambridge*, Cambridge University Libraries. https://www.cam.ac.uk/stories/the-rising-tide; https://www.cam.ac.uk/about-the-university/how-the-university-and-colleges-work/processes/honorary-degrees/selected-honorands

Capel, T., Taylor, J.L., & Vyas, D. (2016). Using self-reported experiences to explore the issues of women in crisis situations. In *Proceedings of the 28th Australian conference on computer-human interaction (OzCHI '16)* (pp. 483–488). Association for Computing Machinery. https://doi.org/10.1145/3010915.3010962

Chew-UK Parliament. (2018). Dr Esyin Chew – Written evidence (AIC0166). AI in the UK: ready, willing and able? The Select Committee on Artificial Intelligence. *The House of the Lord*. https://publications.parliament.uk/pa/ld201719/ldselect/ldai/100/100.pdf; https://data.parliament.uk/writtenevidence/committeeevidence.svc/evidencedocument/artificial-intelligence-committee/artificial-intelligence/written/69675.html

Chowdhury, F.N., Bhattacharya, B.S., Cho, H.-K., Faragasso, A., Gebeshuber, I.C., Ciuperca, E.M., Marinova, G., & Doyle-Kent, M. (2022). Women in STEM: Snapshots from a few Asian countries. *IFAC-PapersOnLine*, *55*(39), 204–209. https://doi.org/10.1016/j.ifacol.2022.12.060

Dutta, D. (2018). Women's discourses of leadership in STEM organizations in Singapore: Negotiating sociocultural and organizational norms. *Management Communication Quarterly*, *32*(2), 233–249.

Elkington, R. (2022). Advanced research training for the biotribology of natural and artificial joints in the 21st century. *BioTrip: EU Horizon 2020*. University of Leeds. https://biotrib.eu/women-in-science-elena-corner-piscopia-the-first-woman-to-graduate

Fatourou, P., Papageorgiou, Y., & Petousi, V. (2019). Women are needed in STEM: European Policies and incentives. *Communications of the ACM*, *62*(4), 52.

Founou, L.L., Yamba, K., Kouamou, V., Asare Yeboah, E.E., Saidy, B., Jawara, L.A., Bah, H., Sambe Ba, B., Aworh, M.K., & Darboe, S. (2023). African women in science and development, bridging the gender gap. *World Development Perspectives*, *31*, 100528. https://doi.org/10.1016/j.wdp.2023.100528

García-Holgado, A., Díaz, A.C., & García-Peñalvo, F. J. (2019). Engaging women into STEM in Latin America: W-STEM project. In *Proceedings of the seventh international conference on technological ecosystems for enhancing multiculturality (TEEM'19)*. Association for Computing Machinery, 232–239. https://doi.org /10.1145/3362789.3362902

García-Peñalvo, F.J., García-Holgado, A., Dominguez, A., & Pascual, J. (2022). *Women in STEM in higher education, good practices of attraction, access and retainment in higher education*. Springer. https://doi.org/10.1007/978-981-19-1552-9

Guernsey, J.H. (1999). *The lady Cornaro: Pride and prodigy of Venice*. College Ave Press.

Harvard. (2024). Women at Harvard University, Harvard University Archives – Research Guide, Harvard Library. https://guides.library.harvard.edu/c.php?g=1108872&p=8085578

Hutchinson, J. (2014). Girls into STEM and Komm mach MINT': English and German approaches to support girls' STEM career-related learning. *Journal of the National Institute for Career Education and Counselling*, *32*(1), 27–34. https://doi.org/10.20856/jnicec.3205

IET. (2024). Over one million women now in STEM occupations but still account for 29% of STEM workforce. *The IET Press*. https://www.theiet.org/media/press-releases/press-releases-2024/press-releases-2024-january-march/8-march-2024-over-one-million-women-now-in-stem-occupations-but-still-account-for-29-of-stem-workforce

Koblitz, A.H. (2016). Life in the fast lane: Arab Women in science and technology. *The Bulletin of Science, Technology & Society*, *36*(2), 107–117.

LGTML. (2022). Let's Go To Mummies' Lab, UM STEM Centre. https://stem.um.edu.my/let-s-go-to-mummie-s-lab

Makarem, Y., & Metcalfe, B. (2023). Microaggressions as a framework for understanding Women's STEM career experiences in Lebanon. *Sex Roles, 89*(3–4), 155–173. https://doi.org/10.1007/s11199-023-01396-4

McKay, D., & Buchanan, G. (2021). Feed the tree: Representation of Australia-based academic women at HCI conferences. In *Proceedings of the 32nd Australian conference on human-computer interaction (OzCHI '20)* (pp. 263–269). Association for Computing Machinery. https://doi.org /10.1145/3441000.3441061

McKinnon, M. (2022). The absence of evidence of the effectiveness of Australian gender equity in STEM initiatives. *The Australian Journal of Social Issues, 57*(1), 202–214. https://onlinelibrary-wiley-com.ezproxy.cardiffmet.ac.uk/doi/full/10.1002/ajs4.142

MIT. (2017). Scene at MIT: Ellen Swallow Richards leads the Women's Laboratory. MIT News Office. https://news.mit.edu/2017/scene-at-mit-ellen-swallow-richards-womens-laboratory-0321; https://innovation.mit.edu/interactive-timeline-women-at-mit/

Mkhize, Z. (2023). Is it transformation or reform? The lived experiences of African women doctoral students in STEM disciplines in South African universities. *Higher Education, 86*(3), 637–659. https://doi.org/10.1007/s10734-022-00918-5

NSA (2018). National STEM Association Malaysia. https://nationalstemmy.com/v2/about-nsa

Oxford. (2020). Women Making History. https://www.ox.ac.uk/about/oxford-people/women-at-oxford https://www.history.ox.ac.uk/article/a-short-history-of-womens-education-at-the-university-of-oxford

Patterson, L., Varadarajan, D. S., Salim, B.S. (2021). Women in STEM/SET: Gender gap research review of the United Arab Emirates (UAE) - a meta-analysis. *Gender in Management, 36*(8), 881–911.

PIE. (2024). PIE I and PIE II programme, funded by British Council. https://pie4stemwomen.wixsite.com/pie4stem/clients

Pilotti, M.A.E., El Alaoui, K., Abdelsalam, H.M., & El-Moussa, O.J. (2024). Understanding STEM and non-STEM female freshmen in the Middle East: A post-pandemic case study. *Cogent Education, 11*(1), 2304365. https://doi.org/10.1080/2331186X.2024.2304365

QS. (2024). QS World University Rankings 2024: Top global universities. https://www.topuniversities.com/world-university-rankings

Quinn, H., & Hayashi, K. (2021). Reentering the workforce is a focal point at the IEEE women in engineering conference. *IEEE Spectrum.* https://spectrum.ieee.org/reentering-the-workforce-is-a-focal-point-at-the-ieee-women-in-engineering-conference

Rhyfeddol, H., & Menywod, D.H. (2023). *Yr Apel / The Appeal: The Remarkable Story of the Welsh Women's Peace Petition 1923–24 Cymru.* Edited by Mathers, J. & Hopwood, M (bilingual edition). https://nation.cymru/culture/review-yr-apel-the-appeal-the-remarkable-story-of-the-welsh-womens-peace-petition https://www.aber.ac.uk/en/news/archive/2023/10/title-267287-en.html

Sakpere, A.B., Ezika, I., & Isafiade, O. (2024). Pilot Study on Likelihood of Adoption of Immersive Technology for Teaching and Learning of STEM based Course in Tertiary Institutions in Africa: A Case Study of Nigeria. In *Proceedings of the 4th African human computer interaction conference (AfriCHI '23)* (pp. 50–53). Association for Computing Machinery. https://doi-org.ezproxy. cardiffmet.ac.uk/10.1145/3628096.3628748

Sidelil, L.T., Spark, C., & Cuthbert, D. (2023). Being in science and at the same time being a woman is difficult': Academic women's experiences of gender inequalities in STEM academia in Ethiopia. *Women's Studies International Forum, 98.* https://doi.org/10.1016/j.wsif.2023.102717

Stanford. (2016). Looking back Early Stanford women, Standford 125. https://125. stanford.edu/early-stanford-women; https://exhibits.stanford.edu/women/feature/ beginnings; https://exhibits.stanford.edu/women/feature/when-the-world-changed-the-impact-of-world-war-ii-on-women-at-stanford

STEM Women. (2023). Women In STEM Statistics: Progress and Challenges.https:// www.stemwomen.com/women-in-stem-statistics-progress-and-challenges

UK Government. (2023). More women to be supported back into STEM jobs in Government-backed training. *Government Equalities Office.* https://www.gov. uk/government/news/more-women-to-be-supported-back-into-stem-jobs-in-government-backed-training

UK Parliament. (2018). AI in the UK: ready, willing and able? The Select Committee on Artificial Intelligence, *The House of the Lord.* https://publications. parliament.uk/pa/ld201719/ldselect/ldai/100/100.pdf

UN (2023). United Nations, UNESCO in Action for Gender Equality, UNESCO. https://www.unesco.org/en/gender-equality

UN. (2024). United Nations - The League of Nations, The Nations United Office at Geneva. https://www.ungeneva.org/en/about/league-of-nations/overview

US Census. (2021). Women Are Nearly Half of U.S. Workforce but Only 27% of STEM Workers, Census of US Government. https://www.census.gov/ library/stories/2021/01/women-making-gains-in-stem-occupations-but-still-underrepresented.html

Varma, R., Falk, J.H., & Dierking, L.D. (2023). 'Challenges and opportunities: Asian Women in science, technology, engineering, and Mathematics. *American Behavioral Scientist, 67*(9), 1063–1073.

WCIA. (2023). Timeline of Key Dates over Women's Peace Petition Centenary, Welsh Centre for International Affairs. https://www.library.wales/peacepetition; https://www.wcia.org.uk/academiheddwch/womens-centenary-timeline; https://www.peoplescollection.wales/collections/1778851

World Economic Forum. (2020). Gender Equality, 3 things to know about women in STEM. *WEF.* https://www.weforum.org/agenda/2020/02/stem-gender-inequality-researchers-bias/

World Economic Forum. (2023). Global gender gap report 2023: Insight report. World Economic Forum. https://www3.weforum.org/docs/WEF_GGGR_2023.pdf

Wright, M. (1959 [2000]). *The sociological imagination.* Oxford University.

Yamada, A. (2023). STEM: Field demand and educational reform in Asia-Pacific countries. *The Oxford handbook of higher education in the Asia pacific region* (pp. 189–209). Oxford Handbooks.

EUREKA STEM Robotics and Artificial Intelligence Initiatives in Wales

2

Chow Siing Sia and Esyin Chew

2.1 INTRODUCTION TO CARDIFF METROPOLITAN UNIVERSITY AND EUREKA ROBOTICS CENTRE FOR STEM

Cardiff Metropolitan University (CMET) was the Times Higher Education UK and Ireland University of the Year in 2021 due to its people-oriented culture, management, and artificial intelligence (AI)-robotics initiatives. Cardiff Met has been named a top three university in Wales in StudentCrowd's Best Universities in Wales 2024 ranking, voted by students (Cardiff Metropolitan University, 2024a). EUREKA Robotics Centre, founded in 2017, is nested within Cardiff Met. It has over 120 robots and

DOI: 10.1201/9781003514626-2

is one of 11 specialist centres in the UK to access cutting-edge robotics facilities, including social and service humanoid robots that always attract government, academic, and industrial visitors. EUREKA represents the meaning of an 'exclamation of satisfaction upon discovering something, solving a problem, and finding a solution with a fun and engaging experience' (Cardiff Metropolitan University, 2024b). The name is inspired by the highest public vantage point in a building in the Southern Hemisphere and the world's tallest residential tower (The Age, 2006): EUREKA's Skydeck 88. It features The Edge – a glass cube with visitors inside, suspended almost 300m (980 ft) above the ground (Herald Sun, 2007), which is a symbolic expression of Cardiff Met's EDGE (ethical, digital, global and entrepreneurial) ethos and practices.

Grounded in machine learning techniques and data analytics capabilities, the EUREKA Robotics Centre educates, researches, and develops robot applications in the real world, from world-class research inquiry and innovation. The centre is profiled by governments, industries, and universities with world class and inclusive facilities and people. Partnering with Wales, England, and global collaborators from companies and universities, the EUREKA Robotics Centre specialises in service and social humanoid robotics, with the below research strands and expertise:

1. HRaaS Hub for Healthcare/Hospitality Robotics as a Service research and innovation.
2. STEAM (science, technology, engineering, arts, mathematics) Hub for STEM with arts/agriculture; an Intelligent Robot (IR) Maker Lab for 3D design and printing of robotic parts. Providing public STEM engagement with a particular focus on engaging underprivileged groups.
3. Autonomous Robotics Lab: applied AI for mobile robots and remotely operated vehicle/autonomous underwater vehicle (ROV/AUV), developing AI for remotely operated and autonomous systems.
4. BioAI-BioEngineering Lab: the interface between biosciences, AI, and robotics, exploring the applications of AI and robotics to biomedical problems.

As one of the flagship research clusters at the Cardiff School of Technologies, EUREKA Robotics (Ethical-Ubiquitous-Robotics driving Economy Knowledge Accelerator for Wales) is the innovative research hub with the initial focus in (1) STEM education for schools in Wales/STEM Ambassador for Wales; (2) hospitality, and (3) healthcare (EUREKA Robotics Lab, 2019). In promoting gender equality in robotics and AI education, fostering 'Eureka' moments among all students regardless of gender is crucial. Encouraging girls to pursue STEM fields by highlighting diverse role models and providing equitable access to resources can lead to more inclusive innovation. Recognising and celebrating their achievements reinforces the importance of diversity in shaping the future of technology. This chapter reports the second research group above, the journey of the STEAM Hub from 2017 to present.

2.2 BACKGROUND ON ROBOTICS AND AI INITIATIVES IN WALES AND ENGLAND

In recent years, robotics and AI have gained significant attention globally as powerful tools for advancing education, innovation, and society at large. However, Microsoft (2017) highlighted the disparities in computer science education: '[I]n a year when China and India each produced 300,000 computer science graduates, the UK produced just 7,000.' Steve Jobs once stated, 'Everyone in this country should learn how to program because it teaches you how to think' (US Department of Education, 2015). However, Jesen Huang suggested that children should no longer be encouraged to learn code but AI, as the transformative technologies can make everyone a programmer (Business Today, 2024; Computing UK, 2024). Chew (2018) asserted that 'AI is perceived as intelligent technologies, from computers to robots that mimic human's intelligence and five senses for learning, analytical reasoning, decision-making, real-life problem solving and companionship.' Since 2017, we have been included as part of the written evidence to lobby the work of policy-makers at UK parliament to embed robotics and AI into education, from pre-school and primary education through to higher education and the healthcare sector in the UK and recommend that the UK Department for Education consider celebrating a nationwide robotics and AI education week (EUREKA Robotics Lab, 2019). UK-RAS has commenced the UK Robotics Week and School Robot Competition since 2019, jointly with Twinkl and robotics research groups across the UK (UK-RAS, 2024). The First Tech Challenge (2024) in the UK and Ireland is the world's largest robotics competition for student-directed learning, culminating in building and competing robots in a global forum, supported by an industry mentor.

The EUREKA STEM Robotics and AI programme stands out for its significant public impact and efforts to promote inclusivity in STEM fields. This chapter explores the multifaceted dimensions of EUREKA STEM Robotics and AI, focusing on its background, public impact, collaborative efforts, and implications for women's participation in STEM. The First Campus (2024) and now Reaching Wider national programme (2018–2024) are funded by the Higher Education Funding Council for Wales (HEFCW), aiming to increase higher education participation from underrepresented groups by raising educational aspirations and skills, and creating innovative study opportunities and learning pathways to higher education. Accelerated by this national programme, many academics have become the STEAM ambassador from many universities in Wales. Another prominent STEM programme in all four nations of the UK is called STEM Learning (2024), supplying nationwide STEM workshops, linking industries and universities to schools and colleges. Their provision includes: teacher CPD (continuing professional development) in STEM subjects, bringing STEM role models into schools as part of the STEM Ambassador Programme or providing bespoke of study opportunities, long-term support/internship for groups of schools in collaboration with companies, our aim is always the same – to improve lives through STEM education. Both efforts increase the awareness of STEM progression routes, creating STEM educational aspiration and potential career opportunities in STEM to engage young people and underrepresented groups.

In Wales, the landscape of robotics and AI education innovation has been steadily evolving. Efforts have been made to ensure that robotics and AI educational initiatives are accessible to all, with a particular focus on engaging girls and women in these fields (Welsh Government, 2019). Aberystwyth Robotics Club (2019) was one of the first in Wales to bring robotics into the public for civil impact. They conduct a week of robotics and host the UK National Robotics Competition through UK-RAS (2024). Over the years, educational institutions and government agencies have spearheaded efforts across the country to provide opportunities for pupils and students of all genders to explore and excel in robotics and AI, aiming to foster a diverse and inclusive learning environment that reflects the aspirations and capabilities of all. One such is institution is the EUREKA Robotics Centre, which specialises in service and social humanoid robots and collaborates with partners from Wales, England, Malaysia, and around the world. The EUREKA Robotics Centre is the newest member of the UK-RAS (2022) and has acted as the STEM Ambassador for Wales since 2018 (STEM Learning, 2024). Supporting See Science Ltd (2018) and the First Campus initiatives (2018) with hundreds of STEM-Robotics workshops, our reflections and efforts in impacting the real world will be presented in the next few sections. Chew (2018) at that time, with foresight, propounded that 'robots is a moving AI agent acts as an extension of mankind (not as a replacement); like a pair of angel's wings to human, or Doraemon's pocket attached to human,' inspired by the novel idea of McLuhan (1994, 2001), 'The extension of Man...the media is the message!'

2.3 2017–2021: EARLY POLICY INFLUENCES OF EUREKA'S STEM LAB

During the period 2017–2020, the EUREKA Robotics Centre embarked on its inaugural venture into developing robotics and AI education initiatives, marking the beginning of its efforts to shape public impact and societal changes in Wales. This period saw the establishment of foundational programmes aimed at introducing pupils, students, and educators to the field of robotics and AI, laying the groundwork for future educational endeavours. We are the first in the nation to bring humanoid robots with '500-Year History of Humanoid Robotics' workshops into primary schools, high schools, colleges, underrepresented communities in the Welsh valleys, and university undergraduate courses and workshops. For example, Nao robots and EZ robots (frontier technology from Japan and Canada at that time) were brought to schools in the Welsh valleys to provide insight on the fascinating 500-year history of humanoid robots as well as the latest technologies including Google Glass, mind readers, and flying drones (Education Committee-UK Parliament, 2018). Underprivileged pupils learnt how to program a humanoid robot to dance or sing and represent an artefact homed at National Museum Cardiff. They had exciting responses and engaging experiences. On a third visit to the university, the pupils were joined by their parents so that they could showcase their work as part of a celebration ceremony.

With sponsorships from industries and the Welsh Government, we conducted robotics competitions, robotics apps workshops, robot design competitions, 3D printing training for robotics parts, and cutting-edge robotics showcases for children, pupils, and the public. We also imported humanoid robots from Canada, Japan/France, China, and Germany to Wales for investigations and customisations. Educational robotics underpinned by pedagogy are designed and piloted in Wales (Chew et al., 2021; Hawkins & Chew, 2022; Hu & Chew, 2021; McVey et al., 2021, 2022; Najm et al., 2023; Saini & Chew, 2022; Yang et al., 2021). Nevertheless, unlike Singapore (GovInsider, 2016), schools in the UK have limited access to expensive humanoid robots such as the popular Nao and Pepper. Chew (2018) suggests that the UK robotics and AI curriculum from primary to higher education is falling behind other countries, especially the five major world robotics market, the key robotics leaders, and drivers – China, Japan, Korea, US, and Germany – due to the inadequate financial resources for the Fourth Industrial Revolution. A place-based (university/school) strategy for education and skills provision is suggested:

1. To continue the four-steps framework for university/schools engagement across Wales that involved both pupils and their parents.
2. To introduce new robotics and AI higher degrees, i.e. MSc in Robotics & AI; BEng-MEng in Robotics Engineering (Cardiff Metropolitan University, 2022, 2023), that will produce highly skilled robotics graduates to feed into the new robotics-related vacancies for job retention in Wales.
3. To implement the prototype of 'The Next Wave of Learning with Humanoid Robot' by embedding educational robotic tutors across partner schools as an extension of science and technology teachers.
4. To offer opportunities for a flexible internship or apprenticeship programme to recruit pupils from schools and university students; to coach young people to take advantage of future opportunities in preparation for a fourth industrialisation revolution (EUREKA, 2024).

We are greatly concerned at the UK's low accessibility to commercialised service or social robots in comparison to other countries such as Japan, Malaysia and China and this has particular ramifications for the national robotics and AI curriculum, skills and real-life experiences.

Chew (2018)

There is an urgent need for better but affordable technology, including AI and robotics, to enhance the overall educational experience in Welsh schools and hospitals to support students, teachers, and healthcare staff, along with improving poor internet access in some areas, and training for teachers and healthcare staff, which was noted by Dr Esyin Chew of Cardiff Met (Senedd, 2021; Thomas, 2020). Experts in the UK also raise concerns about the AI-digital divide with unequal access to robots by pupils of different backgrounds in Wales, despite the greater use of automation and AI in education (Bermingham, 2020). As evidence of its commitment to this mission, the EUREKA Robotics Centre organised a series of robotics and AI workshops, and hands-on learning experiences, engaging participants from diverse backgrounds and age groups. Furthermore, partnerships were forged with local schools, community organisations, and industry stakeholders to ensure the broadest possible reach and impact of these initiatives. Through these concerted efforts, the centre began its journey towards fostering a culture of technological literacy and innovation, with the goal of driving positive societal change in Wales.

2.4 2021–2024: EUREKA'S IMPACT ON STEM EDUCATION

EUREKA STEM Robotics and AI aims to promote robotics and AI education and empower individuals from diverse backgrounds to pursue careers in STEM fields. Through hands-on learning experiences, mentorship programmes, and community outreach initiatives, the programme seeks to inspire the next generation of innovators in Wales. New global partnerships have been initiated and established globally to further enhance and support this kind of educational trend. The network is made up of funding bodies such as the British Council, The Alan Turing Institution, the Welsh Government, and university and higher educational partners. The EUREKA programme addresses the need for an increase in students choosing STEM subjects at primary, secondary, and further education levels; the public's appreciation of science, research, and innovation; increased capability and knowledge within society; the empowerment of underrepresented groups in STEM subjects; and improved employability skills and knowledge development in the field of robotics and AI (UK Parliament, 2018).

The programme also supports and enhances the professional development of teachers and the personal learning of learners by enabling shared equitable access to available resources, and engaging learners' independence and interactivity in the learning processes, including self-evaluations, creativity, and the transfer of learning. It encourages more young people to get excited about robots by creating games and workshops, inspiring schools to work together to raise the standards of teaching and learning not only in computing, but other subjects, too. This initiative aims to considerably increase more open-ended, student-driven scientific investigations, providing authentic learning experiences and developing the range of skills connected with working with robots in the curriculum, which would directly help learners develop wider skills in computational thinking and creativity.

Through the PIE I (Partnership for Innovation in Employability) and PIE II initiatives, the EUREKA Robotics Centre effectively engages women and girls in the fields of robotics and AI, fostering collaboration with a spectrum of partners in Wales and Malaysia. The EUREKA Robotics Centre was honoured with the Alan Turing Public Engagement Grant award (2022). Through robotics and AI roadshows tailored for STEAM-H (science, technology, engineering, arts, mathematics, and healthcare) fields, the project reached students, particularly girls, from ten Welsh schools, minority ethnic groups, and female healthcare professionals. High-quality videos, news items, photos, and social media updates related to all road-shows were showcased on EUREKA Robotics Centre's and Cardiff Met's social media platforms, including Twitter and LinkedIn (EUREKA Robotics Centre, 2023). One key objective was to increase the visibility of female researchers in robotics and AI and stimulate broader interest among women and girls in the STEM field. Furthermore, written evidence docu-menting the outcomes of Turing-funded research and engagement activities was being developed for submission to governmental bodies such as the UK Parliament, the Welsh Government and Senedd, and partners at the Cardiff & Vale Health Board.

Structured interviews and questionnaires were effectively administered across three roadshows, yielding comprehensive data regarding the interests, perceptions, knowledge, and engagement levels of participants in robotics and AI. These data collection efforts were conducted with informed consent and adhered to strict internal ethical best practices. The participants represented a diverse sociodemographic spectrum, including healthcare professionals (three female), students (45 females), and educators (nine female). Results were captured in qualitative data from video interviews, providing nuanced insights into participant experiences and perspectives. Most participants love robots and were inspired by the STEM workshops.

With the generous support of the Alan Turing Institute Public Engagement Grant award, the EUREKA Robotics Centre has successfully established multiple new relationships with schools across South Wales, encouraging underrepresented communities to engage in robotics and AI. Through the development of these relationships, there will be continued engagement and promotion of the importance and value of STEM in these communities, along with the sharing of learnings and best practices through written evidence. Preliminary analysis of the data and feedback captured at the roadshow events indicates a highly positive response. For example, of the children that were asked 'Did the robotics sessions today inspire you to be more interested in STEM subjects?', the vast majority (76%) answered positively. Similarly, almost half (49%) responded that the roadshow had influenced what they con-sidered studying at university, which is significant considering the historically low uptake of engineering and technology subjects at university by women in the UK. Finally, the children expressed an almost universally positive level of engagement, with 98% reporting a positive view of the robots.

The roadshows also established new networks and contacts to catalyse further outreach efforts aimed at informing and educating the public about robotics and AI, including the numerous opportunities, leading to substantial societal ramifications. The new networks formed between the university and Welsh schools have enabled the EUREKA Robotics Centre to conduct additional robotics and AI outreach sessions targeting female pupils and students. Specifically, Rights Fest Cardiff, organised by Child Friendly Cardiff in 2023 (Sia, 2023); Robot EUREKA at Bray Leino Events at Commonwealth Games Business Hub in 2022 (Bray Leino Limited, 2022); and WiDEning STEM: where the Women Are at AI UK 2023 (AIUK, 2023).

2.5 2022–PRESENT: TRANSFORMATION TO STEAM ROBOTICS-AI HUB PUBLIC IMPACT AND SOCIETAL CHANGES IN WALES AND BEYOND

The EUREKA Robotics Centre is undergoing a transformative shift, evolving from a STEM lab into a dynamic STEAM hub. This transition amplifies collaborative innovation, fostering interdisciplinary engagement among students, researchers, and industry partners. The hub serves as a vibrant ecosystem driving advancements at the intersection of science, technology, engineering, arts, and mathematics. It cultivates talents via targeted outreach efforts, high-impact certified STEAM robotics and AI training programmes and workshops, networking opportunities, and empowering individuals from all backgrounds. The centre also advocates for gender equality and inclusivity, promoting STEAM education and raising societal awareness of robotics and AI. In its transformation into a dynamic hub, it not only advances knowledge and innovation, but also champions gender equality and inclusivity, encouraging broader societal involvement in the rapidly evolving fields of robotics and AI. In particular, the EUREKA Robotics Centre successfully obtained funding from the British Council for Gender Equality Partnerships in Indonesia and Pakistan. These continued the PIE programme: (1) PIE III, led by Dr Shadan Khan, aiming to integrate 12 humanoid robots in six Pakistan universities and accelerate women and girls in Pakistan, partnering with founder and leaders of the STEM Women in Pakistan (STEM Women-PK, 2024); (2) the GENIUS programme expands the EUREKA STEM reach to Indonesian universities with the integration of four humanoid robots. Seven academics will visit the

EUREKA Robotics Centre in June 2024 for continued portfolio development in various robotics and AI training. The GENIUS programme aims to promote gender equality to drive economic development and enhance social welfare in Indonesia. It seeks to address the gender disparities that persist in the country that are limiting women's access to higher and further education, employment opportunities, career advancement, and leadership roles in STEM-AI fields.

In addition, the affordable Polygenic-weLLbeing (PLL) Robot Culinary Experience funded by the Global Grants Scheme promotes cultural understanding among diverse communities, including Chinese and Muslim groups in Cardiff. This initiative fosters integration through technology, facilitating culinary competitions and fusion cooking challenges. It encourages cultural exchange, teamwork, and culinary innovation, benefiting staff and students and fostering inclusivity on campus. The advanced AI-powered cooking robot, which was pilot-tested with elderly people between 2022 and 2024 and supported by the Impact Builder Grant, enhances educational experiences and cultural interactions. This project's impact extends beyond the university, potentially serving as a model for other institutions and wider cultural integration efforts (PLL, 2024). The advanced educational robot, Robot Xiaolongbao (model Mini), is currently being introduced in Wales, Malaysia, Pakistan, and China (Chew, 2023; Sia, 2023a; 2023b; UMPSA, 2023). As of July 2024 EUREKA Robotics Centre own over 120 robots and have integrated more than 20 robots in Wales, Malaysia, Pakistan and Indonesia, indicating cutting-edge robots shall be accessible by all people: 'one child one laptop/robot, one school one robot and one hospital one robot'.

The EUREKA Robotics Centre has emerged as a pivotal catalyst for public impact and societal change in Wales through its pioneering efforts in robotics and AI education. Through evidence-based initiatives – such as tailored robotics and AI educational programmes (as illustrated), outreach activities, and community engagement – the centre has played a significant role in fostering a culture of innovation, empowering pupils and students, educators, and healthcare professionals, and promoting inclusivity and diversity within STEM disciplines. By equipping individuals with essential skills and knowledge, the centre not only contributes to Wales's economic growth and competitiveness, but also drives meaningful societal change by fostering a more technologically literate and equitable society. One of the key contributions of the initiative has been its role in enhancing STEM education in schools and communities across Wales. EUREKA STEM Robotics and AI has helped educators integrate robotics into their teaching practices, making STEM subjects more engaging and accessible to students by providing resources, training, and curriculum support.

Moreover, the initiative has catalysed a shift in interest towards STEM careers, particularly among underrepresented groups such as women and minorities. By showcasing and highlighting the real-world applications of robotics, EUREKA STEM Robotics and AI inspired individuals from all backgrounds to pursue careers in STEM fields. Through partnerships with various stakeholders, including educational institutions, industry partners, and community organisations, the programme has facilitated internships and job placements, creating pathways for individuals to secure more meaningful employment and contribute to Wales's growing technology sector. These collaborations have enabled the initiative to leverage resources, expertise, and networks to reach a broader audience and maximise its impact. Of particular significance are the partnerships forged to promote women's employment and empower female participation in STEM fields. Recognising the under-representation of women in STEM, the initiative has actively engaged with organisations and is dedicated to promoting gender diversity and equity in the workforce. Through mentorship programmes, networking events, and targeted outreach efforts, the initiative has sought to break down barriers and create pathways for women to pursue STEM careers. Furthermore, EUREKA STEM Robotics and AI has also prioritised the recruitment and retention of women in leadership positions, ensuring that diverse perspectives are represented.

The efforts of EUREKA STEM Robotics and AI in promoting women's participation in STEM fields hold significant implications for the broader movement towards gender equity and diversity in STEM. By challenging stereotypes, providing role models, and creating supportive environments, the initiative has demonstrated that women have a vital role to play in shaping the future of technology and innovation. Moreover, the experiences and lessons learned from EUREKA STEM Robotics and AI can serve as a blueprint for other initiatives seeking to increase women's representation in STEM fields. By emphasising collaboration, inclusivity, and mentorship, organisations and educational institutions can create an environment where women feel empowered to pursue and thrive in STEM careers. By promoting gender diversity and inclusivity in STEM fields, the programme is contributing to the creation of a more equitable and inclusive society. By harnessing the talents and perspectives of women, Wales can potentially drive innovation, foster economic growth, and address pressing technological challenges facing the world today. The STEM lab at the EUREKA Robotics Centre has received numerous invitations for STEM workshops following the Alan Turing Public Engagement Grant award and for providing robotics social care support in healthcare projects at Llandough Hospital (and Duffryn Ffrwd Manor Care Home in Wales. Additionally, the Secretary of State for Wales visited the EUREKA Robotics Centre on 17 February 2023 (Business News Wales, 2022).

Through the British Council-funded Catalyst grant, the PIE programme has evolved from PIE I and PIE II to a self-sustainable AI Hub, connecting academics, industries, leaders of higher education, and policymakers for women empowerment. The last chapter the book (Chapter 6) will conclude the key results and implications on how to move STEM robotics to the next level for a global impact.

2.6 SUMMARY

The EUREKA STEM Robotics and AI initiative is radically driving public and societal change and fostering inclusivity in STEM education and employment. In total, EUREKA Robotics STEAM Hub has delivered nearly 300 workshops, impacting over 27,000 members of the public and pupils, including women and girls, and underprivileged communities. Through its collaborative efforts, the initiative has inspired individuals from all backgrounds to pursue careers in STEM fields, with a particular emphasis on promoting women's participation and leadership. As Wales continues to embrace innovation and technology, initiatives like this will play a crucial role in shaping a more equitable and inclusive future for all. The new national curriculum with AI and robotics for the UK in all four nations shall be integrated. We do not merely use low-cost Arduino or RaspberryPi kits for schools, which have worked well in the last decade, but aim to evolve to engaging bio-inspired humanoid robotics with modularised and generative AI educational components.

REFERENCES

Aberystwyth Robotics Club. (2019). Aberystwyth Robotics. https://aberrobotics. club/en/workshops/python/#session-1

AIUK (2023, July 13). WiDEning STEM: Where the women are I AI UK 2023 [Video]. YouTube. Retrieved from https://www.youtube.com/watch?v=-NbiqYZ0mLs&ab_ channel=AIUK

Bermingham, R. (2020). Life beyond COVID-19: What are experts concerned about? Horizon Scanning, UK Parliament. https://post.parliament.uk/life-beyond-covid-19-what-are-experts-concerned-about (Dr Esyin Chew, EUREKA Robotics Lab is one of the 350 experts named)

Bray Leino Limited (2022). Commonwealth Games Business Hub. Retrieved from https://www.brayleinoevents.com/case-study/commonwealth-games-business-hub

Business News Wales (2022, March 16). Secretary of State for Wales Visits World-Leading Research Facilities at Cardiff Met University. Retrieved from https://businessnewswales.com/secretary-of-state-for-wales-visits-world-leading-research-facilities-at-cardiff-met-university/

Business Today (2024). NVIDIA CEO: No Need To Learn Coding, Anybody Can Be A Programmer With Technology. https://www.businesstoday.in/bt-tv/video/nvidia-ceo-no-need-to-learn-coding-anybody-can-be-a-programmer-with-technology-419015-2024-02-26

Cardiff Metropolitan University (2022). BEng-Meng in Robotics Engineering https://www.cardiffmet.ac.uk/technologies/courses/Pages/Robotics-Engineering-BEng-MEng-Degree.aspx

Cardiff Metropolitan University (2023). MSc in Robotics & AI. https://www.cardiffmet.ac.uk/study/newstudents/pg/Documents/2023%20Joining%20Information/MSc%20Robotics%20%26%20AI.pdf https://studentconnect.org/courses/cardiff-metropolitan-university-robotics-and-artificial-intelligence-masters-degree-mscpgdpgc

Cardiff Metropolitan University (2024a). News Cardiff Met ranked a top three university in Wales. https://www.cardiffmet.ac.uk/news/Pages/Cardiff-Met-ranked-a-top-three-university-in-Wales.aspx#:~:text=Cardiff%20Metropolitan%20University%20has%20been%20named%20a%20top,reflection%20of%20student%20experience%20in%20universities%20across%20Wales

Cardiff Metropolitan University (2024b). EUREKA Robotics Centre. Retrieved from https://www.cardiffmet.ac.uk/technologies/Pages/EUREKA-Robotics.aspx

Chew, E. (2018). What are the implications of artificial intelligence? In Love and War. https://publications.parliament.uk/pa/ld201719/ldselect/ldai/100/100.pdf https://data.parliament.uk/writtenevidence/committeeevidence.svc/evidencedocument/artificial-intelligence-committee/artificial-intelligence/written/69675.html

Chew, E. (2023). Educational Robot showcasing at THES Award Panel 2023. https://x.com/EsyinChew/status/1732816363847057901

Chew, E., Lee, P.H., & Khan, U.S. (2021) Robot Activist for Interactive Child Rights Education, International Journal of Social Robotics, The International Journal for Social Robotics, https://doi.org/10.1007/s12369-021-00751-3

Computing UK (2024). Don't encourage kids to code, says Nvidia CEO AI threatens the future of coding. https://www.computing.co.uk/news/4179836/dont-encourage-kids-code-nvidia-ceo

Education Committee-UK Parliament (2019). Written Evidence written UK Parliment Education Committee Fourth Industrial Revolution Latest Evidence - EUREKA Robotics Lab http://data.parliament.uk/writtenevidence/committeeevidence.svc/evidencedocument/education-committee/fourth-industrial-revolution/written/92006.html

EUREKA Robotics Centre (2023, March 9). PIE4 Women in STEAM H Roadshow: Robotics and AI for STEAM-H by EUREKA Robotics Centre [Video]. YouTube. https://www.youtube.com/watch?v=k6_YcoViubs&ab_channel=EUREKARoboticsCentre

EUREKA Robotics Lab (2019). Written Evidence, UK Parliment Education Committee Fourth Industrial Revolution Latest Evidence. http://data.parliament.uk/writtenevidence/committeeevidence.svc/evidencedocument/education-committee/fourth-industrial-revolution/written/92006.html (the article is written by Chew. E.)

GovInsider (2016). Social robots to teach kindergarteners in Singapore by Chin. https://govinsider.asia/intl-en/article/social-robots-to-teach-kindergarteners-in-singapore

Hawkins, G., & Chew, E. (2022). TOM: The assistant robotic tutor of musicianship with sound peak beat detection. In: Ab. Nasir, A.F., Ibrahim, A.N., Ishak, I., Mat Yahya, N., Zakaria, M.A., P. P. Abdul Majeed, A. (eds) *Recent Trends in Mechatronics Towards Industry 4.0. Lecture Notes in Electrical Engineering*, vol 730. Springer, Singapore. https://doi.org/10.1007/978-981-33-4597-3_67

Herald Sun (2007). Experience the terrifying Edge at Eureka Tower." *Herald Sun.* news.com.au. 29 April 2007.

Hu, S., & Chew, E. (2021). The Investigation and Novel Trinity Modelling for Museum Robots, TEEM'20: Eighth International Conference on Technological Ecosystems for Enhancing Multiculturality, October 2020, pp. 21–28, ACM Digital Library. https://doi.org/10.1145/3434780.3436541

McLuhan, M. (1994). Understanding Media: The Extensions of Man (1st ed.). New York: McGraw Hill; reissued by MIT Press, 1994, with introduction by Lewis H. Lapham; reissued by Gingko Press, 2003. ISBN 978-1-58423-073-11967.

McLuhan, M. (2001). The Medium Is the Massage: An Inventory of Effects (1st ed.), with Quentin Fiore, produced by Jerome Agel. House. Reissued by Gingko Press, 2001. ISBN 978-1-58423-070-0.

Mcvey, S. M., Chew, E., & Caroll, F. (2021). The Review of Dyslexic Humanoid Robotics for Reinforcement Learning, European Conference on e-Learning, (Oct 2021). DOI:10.34190/EEL.251.132

Mcvey, S.-M., Chew, E., & Carrol, F. (2022). Educational Robotics & Dyslexia: Investigating how Reinforcement Learning in Robotics can be used to help support students with Dyslexia. The 10th edition of the Technological Ecosystems for Enhancing Multiculturality, Salamanca, Spain, 19th-22nd Oct 2022, TEEMS 2022 https://2022.teemconference.eu

Microsoft (2017). AI Written Evidence Volume., UK Parliament Artificial Intelligence Select Committee's Publications. https://www.parliament.uk/documents/lordscommittees/Artificial-Intelligence/AI-Written-Evidence-Volume.pdf

Najm, A., Chew, E., & Bentley, B. (2023). Robot-Assisted Language Education and Speech Therapy for Children with Cleft Lip and Palate. DOI: https://doi.org/10.34190/ecel.22.1.1787

PLL (2024). PLL Robot Culinary Experience. Cardiff Met University Global Engagement: https://www.linkedin.com/feed/update/urn:Li:Activity:7158466031399677952/

Saini, J., & Chew, E. (2022). The role of 3D-technologies in humanoid robotics: A systematic review for 3D-printing in modern social robots. In: Ab. Nasir, et al. (eds) *Recent trends in mechatronics towards Industry 4.0. Lecture notes in electrical engineering*, vol 730. Springer, Singapore. https://doi.org/10.1007/978-981-33-4597-3_26 (Springer, EI and Scopus-indexed).

See Science Ltd. (2018). See Science: educational and enrichment consultancy in Wales (see-science.co.uk)

Senedd. (2021). Beyond the pandemic: what some experts think, Senedd Research, Welsh Parliament. https://research.senedd.wales/research-articles/beyond-the-pandemic-what-some-experts-think

Sia, C. S. (2023). Rights Fest Cardiff. LinkedIn. Retrieved from https://www.linkedin.com/posts/activity-7123340404493594624-Vomo?utm_source=share&utm_medium=member_desktop

Sia, C. S. (2023a). PIE with Robot Xiaolongbao https://www.linkedin.com/posts/chowsiing-sia-338563194_pie-pie2-stem-activity-7077577113351909376-u0Kj/

Sia, C. S. (2023b). Putrajaya STEM Workshop, organised by UiTM. https://x.com/ChowSiingSia/status/1677631376445833222

STEM Learning (2024). STEM Ambassador for Wales. https://www.stem.org.uk/

STEM Women-PK (2024). STEM Women in Pakistan. https://stemwomen.org.pk

The Age (2006). "Eureka tower officially opens." *The Age*. Melbourne. 11 October 2006.

The First Campus (2024). The First Campus and now Reaching Wider. https://reachingwider.ac.uk; firstcampus.org

The First Tech Challenge (2024). https://firstuk.org

Thomas, L. (2020). Universities in Wales' civic response to Covid-19 (EUREKA Robotics Lab): https://uniswales.ac.uk/sites/default/files/2021-10/Universities-in-Wales-Civic-Response-to-Covid-19%20%2818%29.pdf (Dr Esyin Chew's civil statement)

UK Parliament (2018). Fourth Industrial Revolution [Written evidence submitted by Cardiff Metropolitan University]. Retrieved from https://data.parliament.uk/writtenevidence/committeeevidence.svc/evidencedocument/education-committee/fourth-industrial-revolution/written/92006.html

UK-RAS (2022). Cardiff Met is one of the 37th Members of UK-Robotics and Autonomous Systems Network (ukras.org.uk) https://www.ukras.org.uk/members

UK-RAS (2024). The School Robot Competition and UK Festival of Robotics https://www.ukras.org.uk/school-robot-competition/about; https://www.ukras.org.uk/robotics-week

UMPSA (2023). Robot Xiaolong Bao meeting the teachers from 70 districts and Officers of Pahang State Education Department. https://x.com/EsyinChew/status/1693961443606196609

US Department of Education (2015). White House Initiative on Advancing Educational Equity, Excellence, and Economic Opportunity through Historically Black Colleges and Universities, https://sites.ed.gov/whhbcu/multimedia/everyone-in-this-country-should-learn-how-to-program-because-it-teaches-you-how-to-think-steve-jobs/#:~:text=Home-,%E2%80%9CEveryone%20in%20this%20country%20should%20learn%20how%20to%20program%20because,how%20to%20think%E2%80%9D%20%E2%80%93%20Steve%20Jobs

Welsh Government (2019). Delivering Economic Transformation for a Better Future of Work. https://www.gov.wales/sites/default/files/publications/2019-09/delivering-economic-transformation-for-a-better-future-of-work.pdf

Yang, J., Chew, E., & Liu, P. (2021). Service humanoid robotics: A novel interactive system based on bionic-companionship framework, *PeerJ Computer Science*, 7, e674, https://doi.org/10.7717/peerj-cs.674

Industrial Sponsored STEM Star @UiTM

3

Sukreen Hana Herman and Rozita Jailani

3.1 QUADRUPLE HELIX: SMART PARTNERSHIP FOR COMMUNITY ENGAGEMENT

The concept of the university serving society is fundamental and requires full dedication from every member of the university staff. This commitment ensures that the community benefits from the research, teaching, and learning activities conducted within the institution of higher education. This idea aligns with the objectives outlined in the Malaysian Education Development Plan 2015–2025, which emphasises the importance of universities contributing to the community.

The plan emphasises the establishment of a strong connection, forming the core of the university's role in community engagement, through what is known as the Quadruple Helix Engagement model (Figure 3.1). This model highlights the collaboration between the university, government, industry, and community, aiming to bring about positive changes in knowledge, attitudes, skills, and aspirations within the community (Sally, 2015). Ultimately,

DOI: 10.1201/9781003514626-3

Quadruple Helix Framework

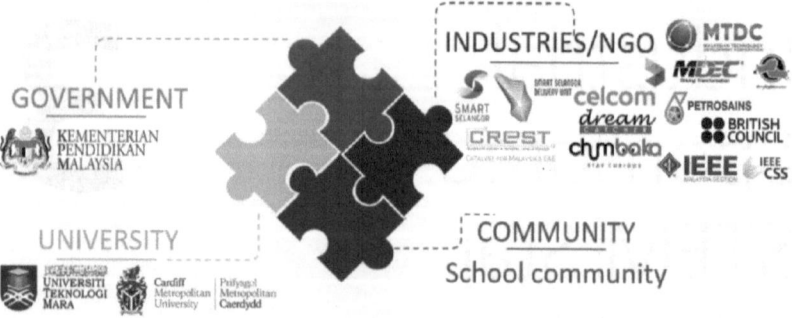

FIGURE 3.1   Quadruple Helix framework.

this model demonstrates the sustainability of the relationships forged between these key stakeholders, paving the way for meaningful societal impact.

Following this model, Universiti Teknologi MARA (UiTM) has encouraged participants in community projects to establish a network and foster close collaboration, ensuring the sustainability of programmes and maximising their impact on beneficiaries. Within this framework, UiTM's academic staff serve as knowledge holders, providing expertise and skills to develop teaching modules tailored to the current needs of beneficiaries. The industry plays a pivotal role as the primary sponsor, providing financial support and meeting programme requirements. Government entities, represented by relevant ministries, departments, offices, or agencies, endorse the programmes. Meanwhile, the community, comprising the beneficiaries, also plays a crucial role in ensuring the successful implementation of planned activities. The Quadruple Helix Engagement model provides a comprehensive framework involving all relevant stakeholders, guaranteeing that community engagement programmes achieve their objectives and generate significant impact.

3.2 STEM STAR @ UiTM: MAKER TALENT FOR DIGITAL INNOVATION

Recognising the critical need for future talent in science, technology, engineering, and mathematics (STEM) fields in Malaysia and understanding the profound impact of STEM disciplines on the economy and national progress, a group of lecturers from UiTM's College of Engineering has taken proactive

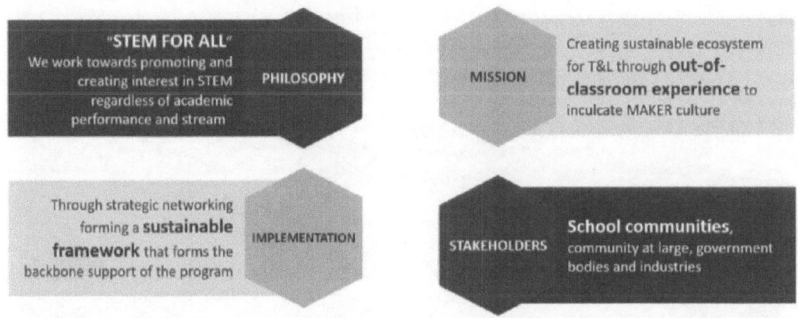

FIGURE 3.2   Philosophy and mission statement of MTDI.

steps to raise awareness and nurture interest in STEM education. This dedicated group, known as Maker Talent for Digital Innovation (MTDI), comprises seven engineering lecturers and one educational technology lecturer from the Faculty of Education.

Embracing the ethos of 'STEM for All,' MTDI is committed to promoting and igniting interest in STEM disciplines regardless of academic performance or specialisation (Figure 3.2). Their mission is to establish a sustainable ecosystem for teaching and learning by fostering hands-on experiences beyond traditional classroom settings, with the aim of instilling a culture of innovation (the MAKER culture).

MTDI firmly believes that this initiative can be effectively realised through strategic networking, creating a robust framework that serves as the cornerstone of the programme's support structure. Such a framework is envisioned to benefit various stakeholders, with a particular focus on the school community, alongside industry, government agencies, and the broader public.

Acknowledging the disparity in STEM awareness and exposure between urban and rural schools, MTDI is dedicated to bridging this gap by providing equal opportunities for all (Benjamin et al., 2022). To achieve this, all MTDI STEM programmes are offered free of charge, ensuring that underprivileged schools receive the same level of exposure as their counterparts.

MTDI's journey commenced with invaluable support from the Malaysia Digital Economy Corporation (MDEC), a government agency established in 1996 under the Ministry of Communications and Digital. MDEC plays a pivotal role in driving Malaysia's digital economy, particularly through initiatives such as the Multimedia Super Corridor (MSC) Malaysia. This initiative aims to establish a nexus for innovative multimedia technology, fostering collaboration between local and foreign entities, government agencies, and ministries to propel Malaysia's socio-economic development in the digital era.

FIGURE 3.3   Digital Maker Hub @ UiTM.

Recognising the significance of cultivating STEM interest among Malaysian teenagers, the MDEC has generously sponsored the creation of Digital Maker Hubs (DMH) nationwide. UiTM is privileged to be selected as one of the higher education institutions to participate in this initiative. Through the DMH programme, UiTM has received sponsorship for the refurbishment of the space and procurement of equipment and furniture, facilitating the establishment of a maker laboratory as shown in Figure 3.3. This laboratory, known as MTDI Space, serves as a hub for various digital maker-based activities.

3.2.1 MTDI Programmes: MDEC ICT Career of Choice

For its inaugural project, MTDI was tasked with providing STEM exposure to school students across Selangor state. MTDI successfully executed the information and communication technology (ICT) Career of Choice (ICT-CoC) programme, which involved training 120 teachers from 12 selected schools. Subsequently, these teachers imparted their knowledge to a total of 6,000 primary and secondary school students.

The ICT-CoC programme, generously sponsored by MDEC, focused on equipping school teachers with foundational skills in electronics prototyping using Arduino microcontrollers. Training sessions were conducted at UiTM's

Digital Maker Hub, where teachers gained hands-on experience. Following their training, each teacher was tasked with training a minimum of 50 students within their respective schools.

Furthermore, students participating in the programme were introduced to STEM and ICT as potential career paths by UiTM lecturers. This holistic approach aimed to inspire and prepare students for future opportunities in these fields.

3.2.2 MTDI Programmes: Young Innovate Programme

The Young Innovate programme is a comprehensive initiative aimed at fostering innovation among students through the use of microcontroller equipment for project development. The programme's implementation involves multiple stakeholders, with DreamCatcher Sdn Bhd and its subsidiary Chumbaka Sdn Bhd serving as the national secretariat, and various industries acting as primary sponsors to fund training activities and provide necessary microcontroller components.

Since its inception in 2016, the Young Innovate programme has made significant strides. By 2019, it had successfully trained 88 schools and 158 students in the utilisation of Arduino microcontrollers and the development of innovation project prototypes. Additionally, the programme engaged UiTM students as mentors, facilitating the collaborative development of innovation projects with school students. Figure 3.4 provides a detailed overview of the Young Innovate Programme's progress and impact from 2016 to 2019.

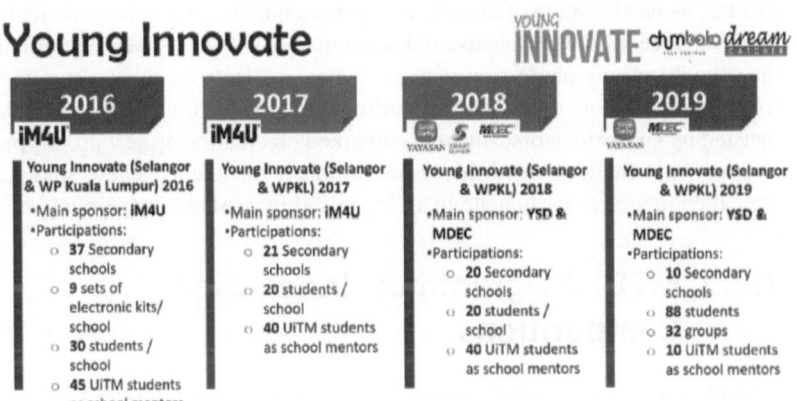

FIGURE 3.4 Young Innovate programme data from 2016 to 2019.

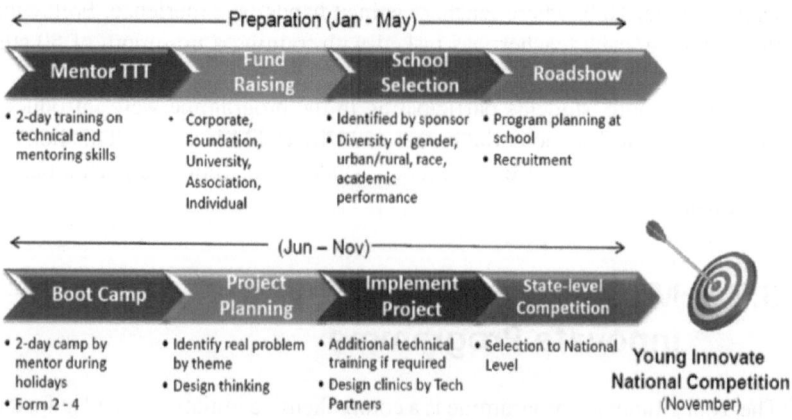

FIGURE 3.5   Young Innovate programme flow.

The Young Innovate programme was structured into several phases, as depicted in Figure 3.5. The initial phase commenced with the Train of Trainers, where UiTM students underwent training and were subsequently appointed as mentors for school pupils. The UiTM students were selected based on their specialisation, predominantly from the electronic engineering programme.

In the second phase, UiTM and the secretariat embarked on seeking sponsorships while simultaneously conducting school selection. Schools were chosen based on their location, distinguishing between urban and rural settings, and considering the number of underprivileged students. Typically, these schools had limited exposure to STEM programmes due to financial constraints, as participation often entails expenses borne by schools or students' parents. Once schools, students, and accompanying teachers were identified, the school training phase commenced. School students received instruction from UiTM mentors under the supervision of UiTM lecturers. Subsequently, assisted by UiTM mentors, students embarked on creating project prototypes aligned with the designated theme and problem statements. These students were then invited to participate in an organised innovation competition.

3.2.3 MTDI Programmes: Innovation Competitions

UiTM has spearheaded a nationwide invention competition to inspire and motivate participants of the Young Innovate programme. This competition, open to all schools, including those not involved in the programme, is provided free of charge. The objective is to offer Young Innovate programme participants

Innovation Competitions

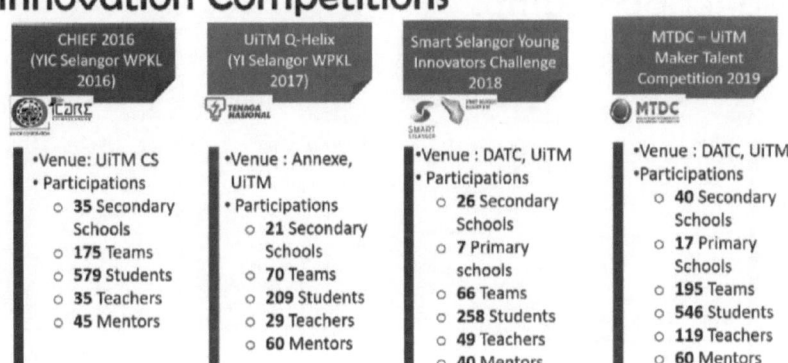

CHIEF 2016 (YIC Selangor WPKL 2016)	UiTM Q-Helix (YI Selangor WPKL 2017)	Smart Selangor Young Innovators Challenge 2018	MTDC – UiTM Maker Talent Competition 2019

•Venue: UiTM CS
• Participations
  ○ 35 Secondary Schools
  ○ 175 Teams
  ○ 579 Students
  ○ 35 Teachers
  ○ 45 Mentors
	•Venue : Annexe, UiTM
	• Participations
	  ○ 21 Secondary Schools
	  ○ 70 Teams
	  ○ 209 Students
	  ○ 29 Teachers
	  ○ 60 Mentors
		•Venue : DATC, UiTM
		• Participations
		  ○ 26 Secondary Schools
		  ○ 7 Primary schools
		  ○ 66 Teams
		  ○ 258 Students
		  ○ 49 Teachers
		  ○ 40 Mentors
			•Venue : DATC, UiTM
			•Participations
			  ○ 40 Secondary Schools
			  ○ 17 Primary Schools
			  ○ 195 Teams
			  ○ 546 Students
			  ○ 119 Teachers
			  ○ 60 Mentors

FIGURE 3.6   Innovation Competitions data organised by MTDI UiTM from 2016 to 2019.

the opportunity to compete alongside schools with extensive experience in STEM innovation competitions, fostering a level playing field. As illustrated in Figure 3.6, the organised competitions were annually sponsored by various industries. Johor Corporation, a subsidiary of the Johor state government, sponsored the competition in 2016, followed by sponsorship from Tenaga Nasional, a Malaysian multinational electricity company, in 2017. In 2018, SSDU Innovations Sdn Bhd, a company owned by the Selangor state government, sponsored the competition, while MTDC sponsored the 2019 edition.

Industry not only served as the primary sponsor, but also played a crucial role as competition jurors. This involvement added value by exposing students to industry professionals' perspectives on their innovative projects, while allowing industry representatives to observe students' talent first-hand. One of the most remarkable aspects of the Young Innovate programme and the innovation competition is the demonstration that underprivileged and rural students can excel when provided with proper guidance, exposure, and training. Their original problem-solving approach, addressing real-world issues they encounter daily, is particularly noteworthy. This underscores the importance of equitable access to education and opportunities for all students, regardless of their background.

3.2.4 MTDI Programmes: Maker Bootcamps

In addition to its long-term programmes, MTDI does also organises Maker Bootcamps sponsored by industry partners (as shown in Figure 3.7). The structure and focus of these bootcamps vary depending on the sponsor, but, generally, they incorporate content provided by UiTM in electronic and

Maker Bootcamps

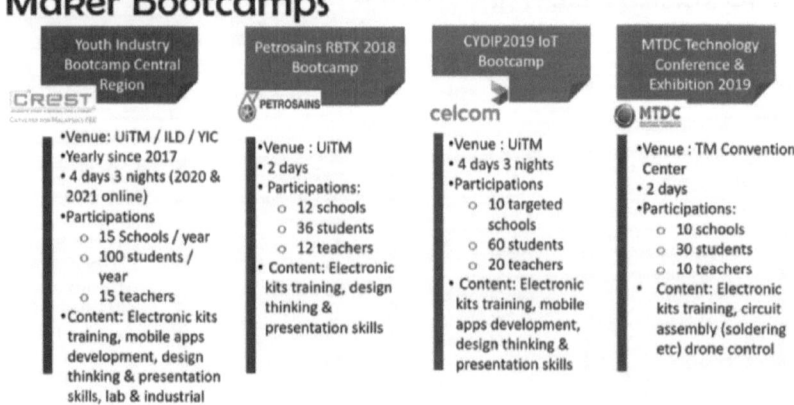

FIGURE 3.7   Maker Bootcamps organised by MTDI UiTM from 2017 to 2019.

robotic engineering, alongside soft skills such as design thinking and presentation skills. Typically, spanning at least two days, the bootcamps commence with participants receiving instruction on microcontrollers using electronic kits. Subsequently, they are presented with a problem statement and tasked with proposing solutions. These proposals are then presented to a panel of industry judges, who offer feedback and advice.

For bootcamps extending beyond two days, participants are not only required to present proposed solutions, but also to construct prototypes of their solutions. These prototypes are showcased to the industry panel, further enriching the learning experience. Maker Bootcamps serve to cultivate students' interest in STEM fields while honing technical and soft skills such as design thinking, problem-solving, and critical thinking. Additionally, they foster resilience by challenging students to complete tasks within a limited timeframe. The collaboration between academia and industry in these bootcamps extends beyond financial support, as industry partners also contribute their expertise to further the shared goal of nurturing STEM talent and fostering interest in the field.

3.2.5 MTDI Programmes: CelcomDigi Young Digital Innovators Programme

The CelcomDigi Young Digital Innovators Programme (CDYDIP) stands as MTDI's flagship initiative, forged in collaboration with CelcomDigi Berhad.

Originating in 2019, this partnership commenced when CelcomDigi Berhad, then known as Celcom Axiata Berhad, sponsored UiTM to launch the Young Digital Innovators programme. The programme focused on training students and teachers from 21 secondary schools in developing digital innovation projects. In 2019, a total of 420 school students and 21 teachers underwent training, resulting in the successful creation of 140 innovation projects. CelcomDigi generously covered all expenses associated with the training, providing laptops, smartphones, and electronic kits directly to the participating schools for the development of prototypes and innovation projects.

The CDYDIP programme comprised training sessions held at MTDI Space at UiTM and at the participating schools, as well as a Maker Bootcamp at UiTM. This bootcamp, tailored for CDYDIP, focused on microcontroller and Internet of Things (IoT) training. Students received comprehensive instruction, from hardware connectivity to mobile app development. During these training sessions and Maker Bootcamps, students were presented with specific themes and encouraged to identify pertinent issues within their surroundings. They were then guided in constructing problem statements and proposing solutions, as well as developing prototypes. Each student group was paired with a mentor from UiTM to facilitate the prototype development process.

Participating schools encompassed a diverse range, including those situated within the interior of Selangor, special education schools for the disabled (PWD), and street children overseen by non-governmental organisations (NGOs). Following their training through the CDYDIP programme, students had the opportunity to compete in innovation competitions alongside a myriad of schools, including international schools, boarding schools, private institutions, and regular day schools. Figure 3.8 depicts some of the activities conducted for this programme.

FIGURE 3.8 Microcontroller and project development trainings conducted at MTDI Space UiTM, sponsored by CelcomDigi Berhad.

Despite the challenges posed by the COVID-19 pandemic, the CDYDIP programme persevered. In 2020, physical activities were rendered impractical due to the unpredictable circumstances, prompting a shift to online engagements. However, in 2021, CDYDIP underwent a significant transformation. Rather than solely focusing on student development, CDYDIP 2021 adopted a holistic approach by equally prioritising the professional development of school teachers.

Introducing the Teacher Webinar Series, CDYDIP 2021 aimed to provide school teachers with exposure and essential training on various applications, software, and digital platforms relevant to their responsibilities. The response from participating teachers was overwhelmingly positive, as they embraced the opportunity to acquire skills necessary for navigating the current technological landscape, which had previously been unfamiliar to them.

The CDYDIP Teacher Webinar Series in 2021 covered five topics and attracted over 4,000 teachers from across Malaysia in live sessions. Furthermore, the recorded sessions on YouTube garnered over twenty thousand views, indicating a widespread interest in the topics covered. The topics addressed in the Webinar Series from 2020 to 2023 are outlined:

2020: Empowering Digital Teaching Techniques

1. 21st Century Learning Design: preparing Teachers for the Millennials.
2. Design Thinking for Sustainable Learning.
3. Creating Creative Online Content.
4. Developing E-Portfolio as Learning Journal.
5. Easy Steps to Create Rubrics.

2021: Preserving Education Through Immersive and Meaningful Learning

1. Embracing Changes through the Science of Learning.
2. Easy Steps of YouTube Videos Making.
3. Best Practices for Designing & Evaluating Open Assessments.
4. Play and Learn Together: gamification 101.
5. Mobile Applications for Education.

2022: Forming Digital Thinking Minds and Innovation in Teaching and Learning

1. Assessing Students Digital Literacy & Strengthening Online Learning.
2. Teaching in the Post-pandemic Era.
3. Developing Thinkers.
4. STEM Education – Way Forward.

2023: AI in Education

1. Decoding Perplexity – Understanding Its Role in Modern AI.
2. Beyond Traditional Tools – Unveiling the Capabilities of Notion AI.
3. Navigating Merlin AI Practical Techniques for Optimised Assessments.

In response to the ongoing pandemic, all sub-programmes under CDYDIP programme transitioned to online platforms starting in 2021. This adjustment included the Maker Bootcamp programme, which was conducted entirely online while retaining its core components.

The online Maker Bootcamp covered a range of topics related to programming, including Python, PHP, SQL, and SCRATCH, providing students with exposure to microcontroller use and programming, website construction, and design thinking. Participation in the bootcamp was open to primary and secondary school students nationwide, ensuring inclusivity and accessibility despite the challenging circumstances posed by the pandemic. Some of the activities carried out are depicted in Figure 3.9.

The CDYDIP innovation competitions persisted in an online format. Participants were tasked with submitting a video presentation along with a brief project description for evaluation during the initial stage. Industry and academic juries reviewed these submissions to select finalists for the subsequent stage. Unlike conventional innovation competitions, CDYDIP's format incorporated a presentation workshop for finalists. This workshop aimed to enhance participants' presentation skills and equip them with the knowledge needed to effectively showcase their projects online during the final round.

Participation in the CDYDIP innovation competition remained open to all, free of charge. Moreover, the competition received full support from the

 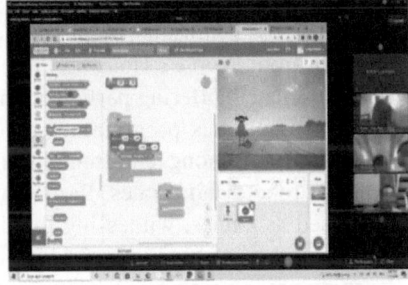

FIGURE 3.9   Online Maker Bootcamps under the CDYDIP programme.

FIGURE 3.10 Online innovation competition under the CDYDIP programme.

Malaysian Ministry of Education, underlining its significance and commitment to fostering innovation and digital literacy among students. CDYDIP innovation competitions also continue to be held online (as shown in Figure 3.10). Competition participants were required to send a video and a brief description of the project to be evaluated in the first stage by industry and academic juries to select the final stage participants. Unlike other typical innovation competitions, the CDYDIP innovation competition includes a presentation workshop for the finalists to help them improve their presentation abilities and learn how to showcase what they have created online at the final round.

3.2.6 MTDI Programmes: DroBotics

An additional collaborative initiative with CelcomDigi Berhad is DroBotics, which commenced in 2021 and has since impacted over 1000 children. In this programme, CelcomDigi generously provided Tello drones and Maqueen line-follower robots for use. DroBotics sessions are conducted at Digital Economy Centers (Pusat Ekonomi Digital (PEDi)) supervised by CelcomDigi Berhad across Peninsular Malaysia. Each session accommodates up to thirty children from People's Housing Projects and rural areas where PEDi facilities are situated, offering participation at no cost.

The DroBotics programme serves to expose and cultivate interest in programming among children through interactive games involving line-follower robots and drones. Participants engage in programming activities disguised as games, witnessing the direct outcomes of their programming efforts (as depicted in Figure 3.11). This approach fosters the perception that learning programming can be enjoyable and straightforward with the right guidance.

FIGURE 3.11   Children trying their line-follower robots during DroBotics.

3.3 STEM STAR @ UiTM: STEM4Fun

The STEM4Fun programme, a joint initiative of UiTM and IEEE Electron Device Society (EDS) Malaysia, aims to cultivate interest and provide exposure to STEM subjects among primary and secondary school students. Through engaging demonstrations featuring electronic blocks (Snap Circuits), drones, green energy technology, and robots, the programme ignites curiosity and enthusiasm for STEM topics. Furthermore, it supports schools in establishing STEM clubs, enhancing early exposure to STEM concepts for students. Partnering with Absidee Group Sdn Bhd and AR Distributor, two prominent companies, the programme aims to inspire children to pursue engineering degrees in disciplines such as mechanics, electrics, and electronics. By adopting the Quadruple Helix Engagement model, involving university entities (UiTM, Universiti Kebangsaan Malaysia, Universiti Sultan Zainal Abidin, and Universiti Malaya), industry partners (Absidee & AR Distributor), NGOs (IEEE EDS Malaysia), and government bodies (Ministry of Education Malaysia, Ministry of Higher Education), the STEM4Fun programme facilitates collaborative efforts to create impactful and sustainable initiatives.

Implemented across states including Selangor, Negeri Sembilan, Terengganu, and Kuala Lumpur, the programme is open to primary and secondary students and teachers of all backgrounds, with priority given to those

FIGURE 3.12   STEM4Fun at schools.

in the B40 category. Teachers specialising in STEM fields also benefit from exposure to teaching aids such as electronic blocks, drones, green energy technology, and robots, empowering them to inspire students to pursue STEM education and careers. To date, the STEM4Fun programme has engaged 1,287 students and 256 teachers from 190 primary and secondary schools (as shown in Figure 3.12). Sponsored by IEEE EDS with a USD 15,000 allocation, the programme covers the purchase of electronic blocks, solar robots, and provides breakfast for students, teachers, and secretariat members, ensuring the programme's success and sustainability.

3.4 CONCLUSION

In conclusion, the Quadruple Helix Engagement model plays a pivotal role in achieving the primary objective of fostering awareness and interest in STEM among young children. Collaboration between universities, industries, government, and related agencies is paramount to effectively implementing planned initiatives and programmes. Close cooperation between academic institutions and industries is crucial for ensuring the sustainability of these programmes. Industries contribute not only financial support, but also invaluable knowledge and expertise, enriching the programmes and initiatives. Industry professionals offer direct feedback and guidance to students and the community, stimulating interest in pursuing STEM degrees and careers. Through this collaborative approach, students are inspired to consider the diverse opportunities within STEM fields, guided by the insights and experiences shared by industry experts. Ultimately, this fosters a pipeline of talent equipped to contribute to and thrive in STEM-related industries, driving innovation and progress in society.

REFERENCES

Benjamin, M., Rachel, P., & Emily, Q. (2022). Words matter: Defining opportunities in STEM to improve rural and urban student outcomes. *Journal of Science Policy and Governance*, *20*(2), https://doi.org/10.38126/jspg200206

Sally, E. (2015). Technology, social innovation, and social entrepreneurship in the quadruple helix. In D.B.A. Mehdi Khosrow-Pour (Ed.), *Encyclopedia of Information Science and Technology* (pp. 2897–2906). IGI Global. doi: 10.4018/978-1-4666-5888-2.CH283

Exploring Challenges, Best Practices, and the Road Ahead for Digital Making Skillsets in STEM Education

4

Nurul Hazlina Noordin

4.1 BACKGROUND: REVOLUTION OF DIGITAL MAKING FOR STEM EDUCATION DEVELOPMENT

John Dewey's 'learning by doing' (1938) educational philosophy has revolutionised 21st century curricula across Malaysia schools with a digital making campaign, the hands-on 2-dimensional (2D) and 3-dimensional (3D) simulation, and coding leading to programmable or intelligent artefacts

DOI: 10.1201/9781003514626-4

(Dewey, 1986). This skillset allows students to mix their technical and creative skills while exploring new ways of bringing computer science to life in the real world, with five main strands: design, programming, physical computing, manufacture, and community and sharing ("Future Learning Digital making curriculum,"). As the world embraces digital transformation, there is a growing emphasis on equipping students with digital making skills to foster creativity, innovation, and problem-solving abilities.

In Malaysia, the integration of digital making into science, technology, engineering, and mathematics (STEM) education began in 2016 with the introduction of subjects including (*Rekabentuk Teknologi* (RT)), fundamental of computer science (*Asas Sains Komputer* (ASK)) and computer science (*Sains Komputer* (SK)) into the national curriculum. The offering of these subjects is due to the increasing demand for programming and digital making skillsets in today's technology-driven job market. This paper elucidates the significance of imparting these skills early in education, addressing their role in fostering essential competencies, promoting creativity, and preparing the younger generation for the challenges of a technology-driven future in the largest state in Malaysia, Pahang.

4.1.1 Digital Making Skillsets

The digital making skillsets are multifaceted and can be broadly categorised into four main domains, namely, programming, design (2D and 3D), physical computing, and manufacturing (Fiore et al., 2021; Gendreau Chakarov et al., 2021; Kafai & Peppler, 2011; Monteiro et al., 2019; Morado et al., 2021; Resnick, 2017; Sun et al., 2022; Vartiainen et al., 2020). The programming domain focuses on coding skills, which encompass the ability to write, modify, and understand software programs. The 3D design domain focuses on students' proficiency in both 2D and 3D design, enabling students to create visually compelling and functional digital and physical prototypes. The physical computing domain is aimed at enabling students to interface digital technology with the physical world, which requires the ability to design and build interactive electronic systems. The fourth domain, manufacturing, is aimed at cultivating the skills needed to fabricate and produce tangible objects, often employing digital tools such as 3D printers and computer numerical control (CNC) machines.

4.1.2 Digital Making Activities in UMP STEM Lab

Established in 2017 within the Faculty of Electrical and Electronics Engineering Technology at Universiti Malaysia Pahang Al-Sultan Abdullah (UMPSA), the

UMP STEM Lab stands as a hub dedicated to enriching STEM skillsets, igniting interest in technology, and bolstering the broader STEM landscape. Positioned at the forefront of STEM education, the lab focuses on initiatives aimed at fostering digital making prowess among school children, teachers, youth, and the wider community.

Built on its core philosophy (encapsulated in Figure 4.1) – 'See, Think, Explore, Marvel' – the UMP STEM Lab encourages curiosity, critical thinking, and exploration. This approach promotes a comprehensive understanding of STEM by urging individuals to observe, analyse, explore, and marvel at the wonders of science and technology. Embracing these principles empowers participants to deepen their understanding, improve problem-solving skills, and develop a lifelong passion for learning.

The lab continually customises teaching and learning curricula for each engagement, led by engineering lecturers and UMPSA students serving as mentors. This approach ensures alignment with participants' needs while providing invaluable mentorship opportunities for UMP students, enhancing their comprehension of engineering principles and communication skills. To date, the UMP STEM Lab has made a positive impact on 12,347 individuals through its diverse outreach initiatives, encompassing robotics classes, coding competitions, and workshops tailored for school children and youth. The programmes include continuing professional development (CPD) programmes designed to support learning and research in best practices for teaching STEM skills.

FIGURE 4.1   UMP STEM Lab logo.

In collaboration with government and non-government agencies, locally and internationally, the UMP STEM Lab delivers STEM outreach programmes focused on digital making skillsets such as programming, coding, physical computing, 3D modelling, and printing. Through a landscape of STEM outreach endeavours, extracurricular educational pathways, and mentorship initiatives, the UMP STEM Lab emerges as a beacon of digital making excellence, poised to nurture the next generation of innovators and problem solvers.

4.1.3 Teaching Programming and Physical Computing in Schools: From Simply Coding to Digital Making

Teaching programming with physical digital making to school children equips them with foundational skills that are rapidly becoming prerequisites for success in various fields. These skills encompass hands-on problem-solving, logical thinking, and computational reasoning. By engaging in coding and digital projects, children learn to break down complex problems into manageable components, fostering analytical skills that extend beyond technology domains (Grover & Pea, 2013). Moreover, it enhances their ability to critically assess information and interact with technology securely (Meyers et al., 2013).

Programming and digital making encourage children to unleash their creativity and innovation. Through hands-on experiences, children can turn their imaginative ideas into tangible creations, fostering a sense of agency and accomplishment (Eguchi, 2014; Garneli et al., 2015; Selwyn, 2009; Webb et al., 2017). As technology continues to shape every facet of society, preparing school children for a technology-driven future is paramount. Teaching programming and digital making ensures that children are not just passive consumers of technology, but active participants who understand its workings (Resnick, 2006).

This knowledge empowers them to adapt to technological changes, engage with emerging technologies, and contribute effectively to the digital economy. Moreover, it cultivates an early interest in STEM fields, potentially steering more students towards careers in these crucial domains (Webb et al., 2017; Gal-Ezer & Stephenson, 2014; Garneli et al., 2015; Grover & Pea, 2013; Hubwieser et al., 2015; Kelleher & Pausch, 2005; Tabarés & Boni, 2023).

4.1.4 Challenges in Teaching Programming and Physical Computing in Schools

However, there are substantial challenges in delivering effective program-ming and digital making education, particularly in certain regions such as the east coast of Malaysia, and in the suburban and rural areas. Teachers in these contexts often face issues that hinder optimal skillset delivery. The digital divide, marked by unequal access to technology and the internet, poses a significant hurdle. Limited resources and outdated infrastructure can inhibit children's exposure to the practical applications of programming and digital creation, thereby impeding their hands-on experiences and creative potential (Garneli et al., 2015; Grover & Pea, 2013)

Educators in these areas struggle with limited access to specialised training and ongoing professional development. The ever-changing nature of technology requires constant skill enhancement, but the lack of train-ing opportunities can hinder teachers from delivering thorough instruction (Gal-Ezer & Stephenson, 2014; Garneli et al., 2015). Additionally, socio-economic factors can exacerbate these challenges, impacting students' readi-ness to engage with programming and digital making education. Addressing these challenges requires a multi-faceted approach, involving collaborations between educational institutions, government bodies, and private sector stakeholders.

4.1.5 Inclusion of Technology Design and Fundamental Computer Science Subjects in Formal Curriculum

The Ministry of Education in Malaysia has taken significant steps to fostering a digital making mindset among schoolchildren, by integrating these skillsets into the formal curriculum. This educational initiative aligns closely with the objectives of technology design (RBT) and fundamental of computer science (ASK) curriculum for primary and secondary schools. The content for both RBT and ASK is included in Table 4.1.

The primary goal of these subjects is to nurture students with essential skills and a positive attitude towards design, entrepreneurship, and technol-ogy. This curriculum places a strong emphasis on adapting to information and communication technology in design, which include the utilisation of digital tools and software. Furthermore, it encourages the application of technology

TABLE 4.1 Comparison of two optional programmes in Malaysian public schools

CONTENT OF RBT (DESIGN TECH)	CONTENT OF ASK (FUNDAMENTAL OF COMPUTER SCIENCE)
The subject RBT has a well-defined objective focusing on cultivating creativity, problem-solving abilities, and a comprehensive understanding of design processes and technology integration.	The subject ASK provides students with a foundational understanding of computer science principles and programming.
Its core aim is to equip students with the ability to conceptualise and construct products and solutions that are functional and visually well-designed.	Its aim is to equip students with computational thinking skills and a comprehensive knowledge of how computers function.
The curriculum places significant emphasis on the complete design cycle of product development, encompassing the stages from idea generation to prototyping and eventual production. It involves guiding students in the processes of problem identification, research, conceptualisation through sketching, and the utilisation of an array of tools, materials, and technologies to actualise their design visions. Furthermore, foundational principles in ergonomics, sustainability, and user-centred design are inculcated.	The curriculum focuses on fundamental concepts in computer science, covering topics like algorithms, data structures, problem-solving strategies, and interactions between computer hardware and software. Students were taught programming languages like Python and Scratch to learn how to write simple code. Through this subject, students learn logical thinking, the ability to break down problems into manageable components, and the skill to develop step-by-step solutions through coding.
This subject supports the development of a diverse skillset, encompassing technical drawing, prototyping, 3D modelling, and hands-on craftsmanship. Additionally, it equips students with proficiency in leveraging digital tools and software for design and manufacturing, notably Computer-Aided Design (CAD) software.	The learning outcomes include the ability to create simple software applications and to gain insight into the role of computers in modern society.
The RBT subject not only prepares students for prospective careers in fields such as industrial design, architecture, engineering, and product development, but also instills skills that are relevant to entrepreneurship and innovation.	The subject ASK is not only valuable for students aspiring to pursue careers in computer science, software development, data science, and related technology fields, but also serves a foundational understanding of how technology shapes our world, making it relevant for all students in the digital age.

in design and product development, which may involve incorporating electronics into projects.

The curriculum aims to equip students with the skills required to effectively employ appropriate tools and materials to create products, which may encompass the use of electronic components. While the curriculum does not explicitly mention coding or electronics design, it is designed to provide students with the holistic knowledge, skills, and values needed to thrive in an ever-evolving technological landscape, which may include these essential skills.

The Ministry of Education's efforts in Malaysia are geared towards nurturing a digital making mindset among school children, emphasising critical thinking, creativity, innovation, sustainability, and responsible technology use. These initiatives aim to prepare students for success in a rapidly changing technological world, as explored in our academic paper.

Both design technology and fundamental of computer science play a pivotal role in the holistic development of well-rounded, technologically proficient individuals who possess the capabilities to innovate and tackle contemporary challenges. These subjects align seamlessly with Malaysia's educational initiatives, effectively preparing students for the dynamic and evolving realms of technology and design.

4.1.6 Contextualising the Community of Inquiry Workshop: Pahang, Malaysia

This study outlines the outcomes of a Community of Inquiry practice (CoI) that provided educators with a platform to deliberate on the challenges, best practices, and future prospect of incorporating digital making and robotics into STEM education (Garrison, 2016). The CoI comprises teaching cognitive and social presence to empower participants, who, in this case, were 70 teachers and teacher training coaches from 65 districts of Pahang, as illustrated in Figure 4.2 (Garrison & Arbaugh, 2007; Swan & Ice, 2010; Swan et al., 2009). Participating teachers' views and discourse were examined. A concurrent triangulation mixed methods design was used with the pre-CoI survey, in-person CoI workshops focus group, international keynote from Wales as a control group input, and post-CoI follow-up.

Situated on the eastern coast of Peninsular Malaysia, Pahang presents a distinctive geographical and educational landscape that warrants examination. In recent years, Pahang has witnessed a notable surge in interest towards

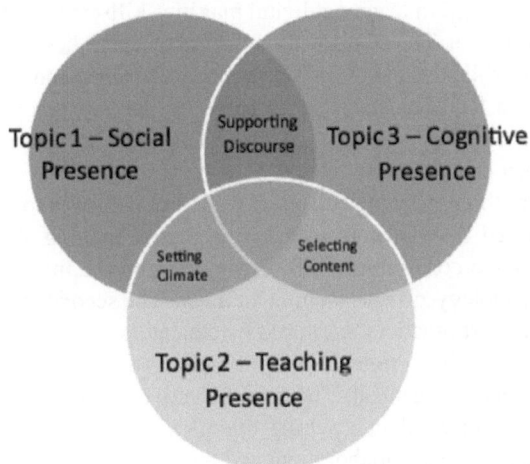

FIGURE 4.2   Community of Inquiry of Pahang Col.

STEM education and digital literacy. This burgeoning enthusiasm aligns with the global recognition of the pivotal role STEM plays in fostering critical thinking, innovation, and problem-solving skills.

As an instrumental part of Malaysia's educational fabric, Pahang's commitment to preparing its students for the digital age is evident in various educational policies and initiatives. Noteworthy among these is the integration of digital skills in curricula, reflecting a proactive approach to equipping students with the competencies essential for navigating the contemporary technological terrain. Against this backdrop, the focus group discussion and survey on 'Empowering the Future: Digital Making Skill Sets in STEM Education in Pahang Malaysia' emerges as a timely endeavour. By engaging educators in robust discussions and practices, the workshop seeks to bridge the gap between theoretical understanding and practical application of digital making skills, aligning with Pahang's aspirations for progressive STEM education in the region.

4.2 RESEARCH METHODOLOGY

This study is guided by specific research objectives and hypotheses aimed at exploring the challenges (teaching presence), best practices (social presence), and the way forward in teaching digital making skillsets in STEM education (cognitive presence). The primary focus of the research lies in understanding the multifaceted landscape of digital making education and the experiences of educators in Pahang, Malaysia. Located on the east coast of Peninsular Malaysia, Pahang has a unique geographical and educational setting that deserves closer examination.

There are three main objectives. The first objective is to investigate the challenges faced by educators in teaching digital making skillsets, with a focus on resource constraints, skill disparities, and the integration of traditional and technology-driven teaching methods. The second objective focuses on identifying best practices employed by educators in teaching digital making skillsets, including strategies for addressing diverse student needs and fostering student engagement. The final objective is aimed at exploring the envisioned way forward in teaching digital making skillsets, considering emerging technologies, collaboration, and the practical application of digital skills in real-world contexts.

4.2.1 Hypotheses

The hypotheses of this study are as follows:

1. Hypothesis 1 (Primary) – The study will reveal that educators encounter various challenges when teaching digital making skillsets in STEM education, including issues related to resources, skill disparities, and the integration of traditional and technology-driven teaching methods.
2. Hypothesis 2 (Secondary) – Educators will adopt individualised teaching approaches to address diverse skill levels and interests among students in digital making education.
3. Hypothesis 3 (Secondary) – Balancing traditional teaching methods with newer, technology-focused approaches will be a common challenge among educators.
4. Hypothesis 4 (Secondary) – Collaborative and innovative strategies will be employed to engage students who may initially be hesitant about digital making.
5. Hypothesis 5 (Secondary) – Professional development opportunities for educators and collaborations with external stakeholders will be highlighted as crucial for enhancing digital making education.

6. Hypothesis 6 (Secondary) – The future of digital making education will involve integrating emerging technologies, fostering collaboration, and emphasising real-world applications of digital skills.

The study's methodology involved focused group discussions, interviews, and surveys to gather insights from educators. Responses were analysed based on their subject matter expertise (SME) as clustered in Figure 4.1(c). Participants were provided with a set of questions corresponding to each topic to facilitate collaborative exploration of ideas, experiences, and strategies.

4.2.2 Qualitative Research Methodology

In addition to the objectives outlined above, this study employed a qualitative research methodology so as to gain a deeper understanding of educators' experiences and perceptions regarding digital making education in Pahang. A total of 73 teachers across Pahang participated in the study. The qualitative research component of this study included focused group discussions, interviews, and surveys to gather qualitative data.

Focused group discussions were conducted with a select group of educators to facilitate collaborative exploration of ideas, experiences, and strategies related to teaching digital making skillsets. These discussions provided a platform for participants to share their views and engage in in-depth conversations.

The interviews were conducted with a subset of educators to delve deeper into their personal experiences, challenges, and innovative practices in digital making education. These interviews allowed for a more personalised and in-depth exploration of the research topics.

Surveys were administered to a larger group of educators to collect structured data on specific aspects of digital making education. These surveys complemented the qualitative data by providing quantitative insights into certain trends and patterns.

Responses from all qualitative research methods were analysed based on participants' subject matter expertise (SME) as clustered in Figure 4.3. Participants were provided with a set of questions corresponding to each topic to ensure a comprehensive exploration of their experiences and perspectives. The qualitative data gathered through these methods enriched the overall findings of the study, adding depth and context to the research objectives and hypotheses.

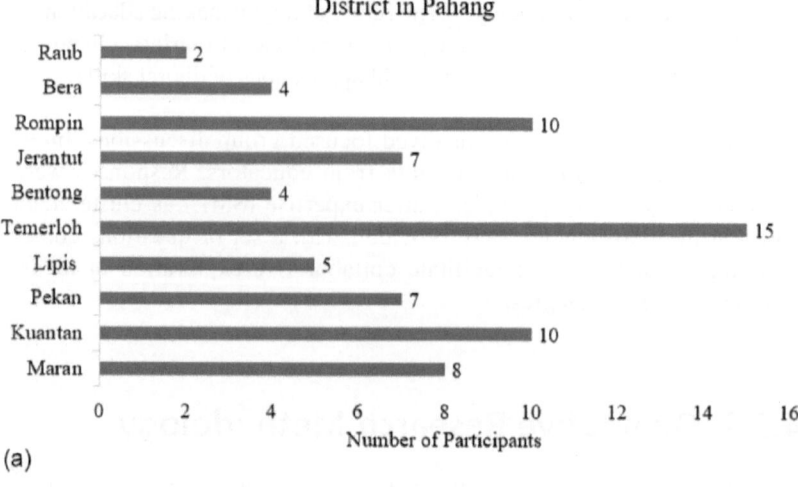

(a)

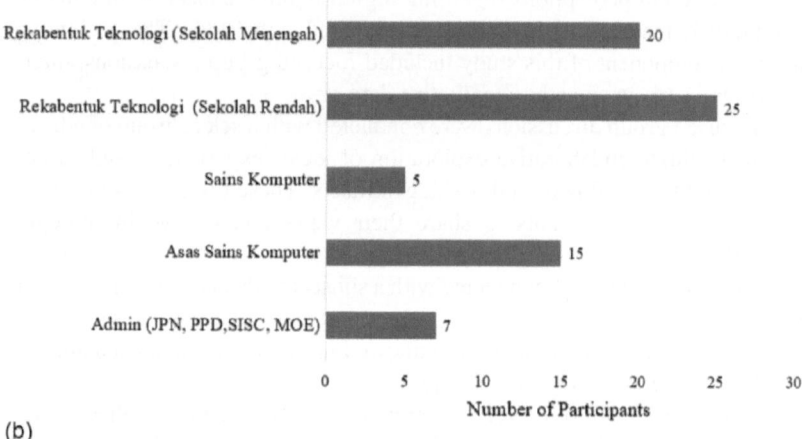

(b)

FIGURE 4.3   Demographic of the participants based on (a) location – districts in Pahang, (b) subject taught in class, (c) age group, and (d) experience in teaching the subjects. *(Continued)*

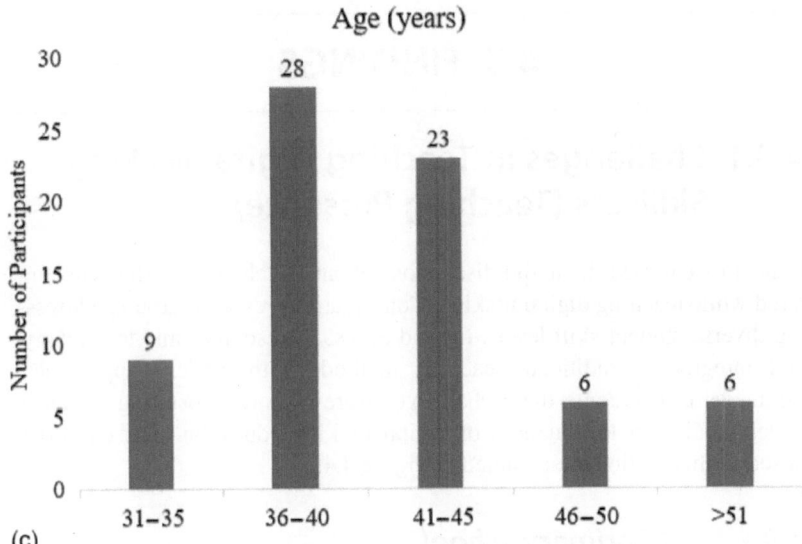

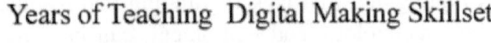

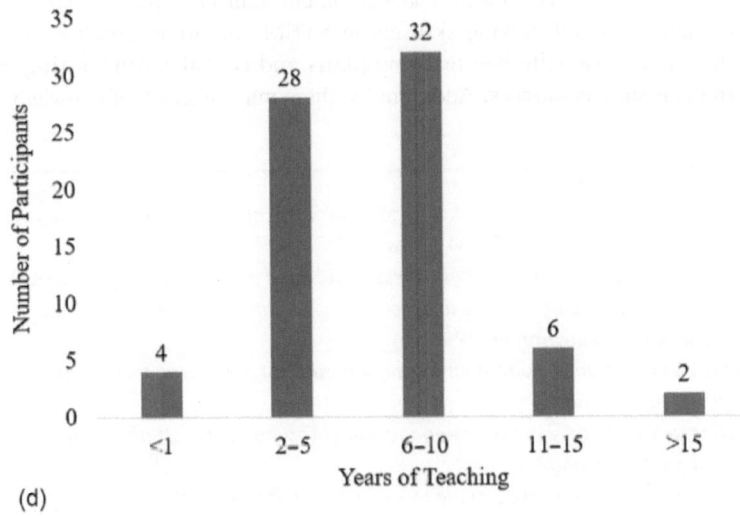

FIGURE 4.3   *(Continued)*

4.3 FINDINGS

4.3.1 Challenges in Teaching Digital Making Skillsets (Teaching Presence)

Educators engaged in candid discussions about the challenges they encountered while teaching digital making. Conversations revolved around addressing diverse student skill levels, limited access to resources and technology, and integrating traditional teaching methods with modern approaches. Strategies to overcome these challenges were explored, fostering a deeper understanding of the nuances of digital making education. The questions posed in this section are as stated in Figure 4.4.

4.3.1.1 RBT primary school

The respondents highlighted that lack of funding and resources, including access to necessary technology and equipment, can be a major challenge when teaching digital making skillsets in STEM education. Teachers may need to get creative with their teaching **plans and consider fundraising to obtain necessary resources.** Additionally, there may be a lack of knowledge

---

1. *What are some of the major challenges you've encountered when teaching digital making skillsets in STEM education?*
2. *How do you address the diversity of skill levels and interests among students?*
3. *Are there any specific barriers to accessing necessary resources or technology in your teaching environment?*
4. *How do you balance traditional teaching methods with newer, technology-focused approaches?*
5. *What strategies do you use to engage students who might initially be hesitant about digital making?*
6. *How can digital making be made more inclusive and accessible to underrepresented groups in STEM?*
7. *Share a specific example of a challenge you faced while teaching digital making and how you overcame it.*

FIGURE 4.4 Questions formulated to discuss the challenges in teaching digital making skillsets.

or skills in specific areas among the teachers, such as programming, which requires a certain level of mastery before they can effectively teach their students.

To address the diversity of skill levels and interests among students, teachers adopted an individual approach method that considers the different needs of each student. For example, for RBT subjects, where coding and technology are dominant, there is a tendency for students to have a higher level of interest in the subject matter due to its hands-on approach. This, however, is subjected to individual differences in learning styles and interests.

Specific barriers to accessing necessary resources or technology in the teaching environment include a lack of appropriate devices, poor learning spaces at home, and higher stress level among students. Additionally, there may be a lack of fieldwork and other hands-on experiences that are necessary for effective learning.

Balancing traditional teaching methods with newer, technology-focused approaches is also a challenge among the RBT primary school teachers. Teachers may need to use a blended teaching approach that combines both methods to create a more effective learning environment.

A specific example of a challenge faced while teaching digital making could be a lack of necessary resources, such as micro-bit kits (microcontroller) or other equipment. An initiative to address this challenge is by adopting online simulator and open-source kits.

4.3.1.2 RBT secondary school

The responses provided highlight several challenges faced by teachers when teaching digital making skillsets in STEM education. These challenges include a lack of funding, inadequate resources, limited access to technology, and insufficient teaching materials. Teachers also face difficulties in addressing the diversity of skill levels and interests among students, balancing traditional teaching methods with newer, technology-focused approaches, and engaging students who may initially be hesitant about digital making.

One specific example of a challenge faced by a teacher was the difficulty in obtaining the necessary components for a mechatronics project for each student. To address this issue, the teachers made a schedule for sharing components among students. Infrastructure wise, challenges faced by the students taking this subject include a lack of appropriate devices, poor learning space at home, and stress. Similar to the previous responses, competencies among the teachers are a major factor, especially in adapting to the current tools and techniques.

Teachers creatively adopted various strategies in engaging students with low interests in the classroom, such as showing examples of successful projects, using technology-based teaching aids, and introducing adaptive technologies that make learning accessible for all students.

To make digital making more inclusive and accessible to underrepresented groups in STEM, teachers can use an individual approach method that takes into account the different needs of each student. Additionally, teachers can introduce adaptive technologies that make learning accessible for all students and create opportunities for all students to participate.

4.3.1.3 Asas Sains Komputer

The respondents highlighted several challenges when teaching the subject. The rapid pace of technological change was identified as a major hurdle, rendering certain content in textbooks outdated and not reflective of current advancements. Challenges are also found in terms of infrastructure, as they experience limiting internet speeds that hindered the accessibility of digital resources and real-time online engagement, affecting the quality of instruction.

Various strategies were implemented to tackle the diversity of skill levels among students. These strategies included tailoring question formats to cater to different student abilities, introducing gamification elements to engage students of various skill levels, and establishing a peer mentor system to bridge the gap between proficient and struggling students.

Teachers acknowledged specific barriers in accessing necessary resources or technology within their teaching environments. These challenges included inadequate laptop availability for students, limited access to high-speed internet, and a scarcity of digital devices for effective learning. However, educators adapted to these constraints by adjusting their teaching methods to suit the available resources.

Upon addressing the topic of traditional and modern teaching methods, finding a balance between traditional teaching methods and newer, technology-focused approaches was a point of emphasis. The teacher discussed their methods of referring to content from both traditional textbooks and modern sources. Additionally, they emphasised the importance of practical demonstrations to showcase the capabilities of new technologies to students.

The teachers shared their strategies to engage students initially hesitant about digital learning including capitalising on peer influence, encouraging participation in external workshops and competitions, and integrating relatable real-world scenarios to spark interest and curiosity among students.

The concept of inclusivity in digital making education emerged as a critical concern. The focused group discussed integrating computational thinking principles across various learning levels to ensure all students could grasp the concepts effectively. They also mentioned employing gamification techniques to accommodate diverse skill levels and learning preferences.

The teachers also highlighted examples of specific challenges faced during teaching and the ways they addressed them. These solutions ranged from embracing cloud technology to mitigate limitations of hardware availability, to using social media groups for sharing among students, and creatively leveraging available offline resources to enhance the learning experience.

4.3.1.4 Sains Komputer (Computer Science)

When teaching *Sains Komputer*, teachers faced challenges including their lack of confidence in programming, hindering the facilitation and guidance towards the development of applications and coding projects. The teachers diversify their teaching methods by combining traditional approaches with newer technology-focused techniques. Despite these efforts, certain barriers persist, such as limited hardware resources and unreliable internet connectivity. In response, teachers creatively use the available resources and adapt their teaching methods to suit the given environment.

To strike a balance between traditional teaching methods and technology-focused approaches, the teachers combined time-tested practices with innovative technology-based tools, as to create well-rounded and engaging lessons. To engage students who may initially be hesitant about learning technology, educators employ diverse strategies. These include designing intriguing and interactive activities, providing clear explanations, promoting collaborative projects, and even serving as role models to ignite students' curiosity.

The teachers made efforts to collaborate with more knowledgeable peers, attend relevant courses, and leverage available technology to enhance their own understanding and, consequently, their teaching capabilities. Similarly, hardware limitations and students' difficulty in staying focused during activities present challenges. In response, educators encourage collaborative efforts and utilise interactive tools to maintain engagement.

In the evolving landscape of technology education, educators remain dedicated to overcoming challenges through creative pedagogical approaches. By integrating various teaching techniques, adapting to resource constraints, and fostering collaborative learning environments, they strive to ensure meaningful and effective technology education for their students.

4.3.2 Best Practices in Teaching Digital Making Skillsets (Social Presence)

This session showcased educators' successful teaching methods that effectively integrated digital making skillsets such as programming and 3D modelling into curriculum. The incorporation of real-world applications and problem-solving strategies was discussed, underscoring the practicality of the skills being taught. Collaborative learning, teamwork, and innovative assessment methods were highlighted as crucial elements in enhancing the learning experience. The questions posed in this section are as stated in Figure 4.5.

The provided responses outline various perspectives on the impact of emerging technologies and trends on education, focusing on digital making and related subjects. Virtual reality (VR), augmented reality (AR), and platforms like TikTok were mentioned as significant emerging technologies that can impact education. These technologies offer opportunities to empower educators, enhance student engagement through competitions, and provide immersive learning experiences. The responses emphasise the importance of preparing students for real-world applications incorporating digital skills and fostering a generation of innovators.

The educators also highlighted artificial intelligence (AI), robotics, blockchain, drones, and Internet of Things (IoT). These advancements hold tremendous potential to revolutionise the way students learn and engage with the world around them. Notably, AI, robotics, and digital making are recognised as pivotal tools that can empower students to not only acquire new skills, but also to become creators and innovators. The vision extends beyond

---

What teaching methods or approaches have you found most effective in fostering digital making skills?

Can you share a successful project or activity that effectively engaged students in digital making?

How do you incorporate real-world applications and problem-solving into your digital making lessons?

What role does collaboration or teamwork play in enhancing the learning experience for digital making?

How do you assess and provide feedback on students' progress in digital making?

Share strategies for adapting your teaching to accommodate students with different learning styles.

How can interdisciplinary approaches be integrated into teaching digital making in STEM education?

FIGURE 4.5   Questions formulated to discuss the Best Practices in Teaching Digital Making Skill.

traditional education, aiming to foster a generation of individuals who possess the confidence and capabilities to tackle complex technological challenges. By integrating these technologies into educational settings, the goal is to ignite a passion for exploration, problem-solving, and creative thinking, ultimately equipping students with the skills needed to thrive in an increasingly digital and interconnected world.

Professional development opportunities are highlighted as essential for educators to keep up with new technologies. Collaboration with commercial companies and other stakeholders is suggested to facilitate skill development and innovative teaching methods. The importance of partnerships and collaborations with industry, higher education institutions, and local communities is emphasised to enhance students' learning experiences by providing exposure to real-world applications of digital making.

To effectively teach digital making, schools should ensure they have adequate infrastructure, including hardware and secure networks. Collaboration with external entities and investment in technology are mentioned as strategies to achieve this. The application of digital making skills beyond the classroom is emphasised, including collaborations, real-world projects, and competitions. These activities aim to prepare students for practical scenarios and lifelong learning.

The vision for an ideal future involves integrating emerging technologies like AI, VR, and AR into STEM education especially in digital making, fostering hands-on learning, collaborative projects, and problem-solving to create a generation of innovative thinkers.

These responses highlight the importance of integrating emerging technologies into education, fostering collaboration and partnerships, and preparing students for real-world applications of digital skills.

4.3.3 Way Forward in Teaching Digital Making Skillsets (Cognitive Presence)

In this final session, educators engaged in forward-looking discussions, anticipating the future landscape of digital making education. The impact of emerging technologies on teaching methods was deliberated, offering insights into adapting to evolving trends. The importance of professional development for educators and the potential of partnerships to enhance students' digital making experiences were explored. The need for schools to provide essential infrastructure and resources for effective digital making education was emphasised, along with suggestions for practical applications of digital making skills beyond the classroom.

The questions posed in this section are as stated in Figure 4.6.

1. *What emerging technologies or trends do you think will have a significant impact on the future of digital making in education?*
2. *How can professional development opportunities better support educators in developing their digital making skillsets?*
3. *What partnerships or collaborations can be established to enhance the digital making learning experience for students?*
4. *How can schools ensure they have the necessary infrastructure and resources to effectively teach digital making?*
5. *In what ways can digital making skills be applied beyond the classroom, and how can we prepare students for those real-world contexts?*
6. *Share your vision for an ideal future where digital making is fully integrated into STEM education.*

FIGURE 4.6   Questions formulated to discuss way forward.

The educators emphasise the importance of integrating emerging technologies like AI, robotics, virtual reality, and coding into educational practices. Collaboration is seen as essential, not only between educators and students but also with external experts, industry partners, higher education institutions, and community organisations. Collaborative efforts can provide valuable resources, expertise, and support to enhance the digital making experience for students.

The need for CPD for educators is highlighted. This includes workshops, training sessions, and collaboration with experts to keep teachers updated on the latest technology trends and tools. Building the digital skills of educators enables them to effectively guide students in their digital making endeavours. To enable effective digital making education, there's an emphasis on having sufficient and up-to-date infrastructure and resources in schools. This includes providing necessary hardware, software, and tools such as 3D printers, robotics kits, and digital content creation platforms.

Responses point towards fostering student engagement through hands-on activities, competitions, exhibitions, and real-world projects. Encouraging students to participate in STEM-related competitions and creating opportunities for them to apply digital making skills beyond the classroom can enhance their learning experiences.

Majority of the participants recognise that digital making promotes creativity and problem-solving skills among students. By engaging in projects that require designing, coding, and creating real-world solutions, students can develop critical thinking abilities and innovation-driven mindsets. The concept of personalised learning is implied in the responses, where educators aim to tailor digital making experiences to suit individual students' interests and needs. Making learning relevant and applicable to real-world contexts is considered essential for preparing students for future challenges.

Emerging technologies facilitate global connectivity and collaboration. Virtual collaboration, online sharing, and communication platforms are seen as opportunities for students to connect with peers from other regions, share ideas, and collaborate on projects. Participants highlighted the importance of preparing students for careers in technology-related fields. Digital making skills are recognised as assets for future employability, entrepreneurship, and addressing industry demands.

Integrating digital making skills across various subjects and co-curricular activities is suggested. Many participants emphasise the relevance of incorporating digital skills in subjects beyond STEM, such as art, social studies, and language arts.

4.4 DISCUSSIONS

4.4.1 Analysis of Hypotheses

The insights gained from the responses of this study have provided valuable key takeaways for advancing digital making skillsets within the context of Pahang's education system. One of the recurring challenges highlighted by participants is the scarcity of essential resources, ranging from funding and materials to necessary equipment and reference materials. Addressing these resource gaps is important especially in enabling educators to effectively teach digital making skills within STEM education, allowing students to engage in hands-on learning experiences that nurture creativity and innovation.

Another critical takeaway revolves around the empowerment of teachers. Many responses highlighted the importance of enhancing educators' knowledge and proficiency in digital making and technology. To overcome this challenge, participants recommended the establishment of regular professional development and training programmes that equip teachers with the necessary expertise to deliver impactful instruction.

In an effort to engage students effectively, various strategies were deployed. These include integrating hands-on activities, project-based learning, collaborative group work, and the incorporation of technology tools such as robotics, 3D printing, and coding. Such approaches aim to ignite students' interest, encourage active participation, and nurture their problem-solving skills. Collaboration emerged as a key theme for advancing digital making skillsets in Pahang's STEM education landscape, in particular the digital making domain. Participants recognised the value of partnerships between educators, schools, institutions, and industry players. These collaborations can provide students with access to valuable resources, specialised expertise, and unique opportunities that enrich their learning journey.

Balancing traditional teaching methods with emerging technologies was also highlighted as essential. Educators are encouraged to integrate cutting-edge tools like AI, VR, and robotics into the curriculum. This fusion promises enhanced learning experiences that prepare students to navigate the demands of an ever-evolving technological landscape. Ensuring equitable access to technology emerged as a challenge, with limited internet connectivity and device availability being noted. To bridge this gap, efforts to provide broader access to technology must be pursued, guaranteeing that all students have equal opportunities for learning and growth.

To empower educators and maximise their impact, robust support and professional development opportunities are deemed essential. By equipping teachers with the skills and knowledge required for effective digital making instruction, Pahang can lay a strong foundation for nurturing innovative and tech-savvy individuals. Central to the workshop's discussions was the concept of student-centric learning. Encouraging active participation, fostering critical thinking, and enabling practical application of knowledge and skills form integral aspects of this approach, enriching the overall learning experience.

Finally, securing adequate funding and resources is emphasised as crucial. Investment in hardware, software, materials, and infrastructure is vital for the successful integration of digital making skillsets within STEM education.

The insights garnered from the workshop emphasise the significance of collaborative efforts, resource allocation, and professional development in propelling digital making skillsets within Pahang's STEM education. By fostering teacher expertise, promoting student-centred learning, and embracing emerging technologies, Pahang can create a dynamic educational environment that equips its students with the skills and mindset needed to thrive in a technology-driven world.

4.5 CONCLUSION

The study serves as a CoI, an inclusive platform for educators to discuss challenges, share best practices, and envision the future of digital making education. The insights gained from the workshop highlight the need for ongoing professional development, collaborative learning, and the integration of practical applications in STEM education. The research is focused on the dedicated educators in the region of Pahang, particularly those involved in the ASK (fundamental of computer science) and RBT (technology design) teachings. Through their valuable insights and experiences, this study explores the intricate landscape of digital making in STEM education within this specific context. The challenges these educators face when imparting digital making skillsets were examined. From resource constraints to varying student skill levels and the delicate balance of incorporating traditional and technology-driven teaching methods, profound insights into the unique hurdles encountered in this educational domain were gained. The resourcefulness and adaptability displayed by educators in Pahang, particularly within the ASK and RBT programmes, shone through as they navigated these challenges.

Transitioning to best practices, innovative strategies were employed by the educators in Pahang to integrate digital making education into the curriculum. Whether through real-world applications, problem-solving approaches, collaborative learning, or the use of adaptive technologies, these educators demonstrated a deep commitment to enhancing the learning journey for their students. The emphasis on preparing students for technology-related careers and nurturing a generation of creative problem solvers resonates strongly within the local educational community. Educators in Pahang, especially those engaged in ASK and RBT programmes, collectively recognised the importance of collaboration, not just among students, but also among the educators. This collaborative effort encourages knowledge exchange, innovative teaching methods, and collective growth in digital literacy for the future development of digital making in STEM education.

Future recommendations include a human-centred approach to seek the potential of adapting 'AI-robotics for all' in the curriculum, grounded in pedagogical good practices globally, with the support of the local Pahang State government. Teachers can take advantage of the digital making skillsets in the AI-robotics STEM revolution and access its fruits (UNESCO, 2023), in terms of the future partnership with a global STEM expert for widening access and impact. To maximise the educational effectiveness of digital making skillsets in Malaysia, the role of teachers' coaches/facilitators or expert peers is vital yet sometimes under-played. In addition, national support of the self-sustainable facilities and equipment, teachers' abilities in the thoughtful integration of digital making AI-robotics underpinned by pedagogical good practices are of the vital importance in teacher training and CPD. Gender inequality issues in STEM interest and unbalanced development in STEM education across Pahang regions and Malaysia urban-villages need further international collaborative work. A summary of strategic suggestions is made for the State Government of Pahang, policymakers, teachers, and teacher educators in Malaysia.

REFERENCES

Dewey, J. (1986). *Experience and education.* Paper presented at The educational forum.

Eguchi, A. (2014). Educational robotics for promoting 21st century skills. *Journal of Automation, Mobile Robotics and Intelligent Systems, 8*(1), 5–11.

Fiore, F., Montresor, A., & Marchese, M. (2021). *A Maker Approach for The Future of Learning.* Paper presented at the FabLearn Europe/MakeEd 2021-An International Conference on Computing, Design and Making in Education.

Future Learning Digital making curriculum. (n.d.). Retrieved from https://www.futurelearn.com/info/courses/build-a-makerspace/0/steps/39452

Gal-Ezer, J., & Stephenson, C. (2014). A tale of two countries: Successes and challenges in K-12 computer science education in Israel and the United States. *ACM Transactions on Computing Education (TOCE), 14*(2), 1–18.

Garneli, V., Giannakos, M.N., & Chorianopoulos, K. (2015). *Computing education in K-12 schools: A review of the literature.* Paper presented at the 2015 IEEE Global Engineering Education Conference (EDUCON).

Garrison, D.R. (2016). *E-learning in the 21st century: A community of inquiry framework for research and practice.* Taylor & Francis.

Garrison, D.R., & Arbaugh, J.B. (2007). Researching the community of inquiry framework: Review, issues, and future directions. *The Internet and Higher Education, 10*(3), 157–172. https://doi.org/10.1016/j.iheduc.2007.04.001

Gendreau Chakarov, A., Biddy, Q., Hennessy Elliott, C., & Recker, M. (2021). The data sensor hub (DaSH): A physical computing system to support middle school inquiry science instruction. *Sensors, 21*(18), 6243.

Grover, S., & Pea, R. (2013). Computational thinking in K–12: A review of the state of the field. *Educational Researcher, 42*(1), 38–43.

Hubwieser, P., Giannakos, M.N., Berges, M., Brinda, T., Diethelm, I., Magenheim, J., Pal, Y., Jackova, J., & Jasute, E. (2015). A global snapshot of computer science education in K-12 schools *Proceedings of the 2015 ITiCSE on working group reports* (pp. 65–83).

Kafai, Y.B., & Peppler, K.A. (2011). Youth, technology, and DIY: Developing participatory competencies in creative media production. *Review of Research in Education, 35*(1), 89–119.

Kelleher, C., & Pausch, R. (2005). Lowering the barriers to programming: A taxonomy of programming environments and languages for novice programmers. *ACM Computing Surveys, 37*(2), 83–137. https://doi.org/10.1145/1089733.1089734.

Meyers, E.M., Erickson, I., & Small, R.V. (2013). Digital literacy and informal learning environments: An introduction. *Learning, Media and Technology, 38*(4), 355–367.

Monteiro, A.F., Miranda-Pinto, M., Osório, A.J., & Araújo, C. (2019). Curricular integration of computational thinking, programming and robotics in basic education: A proposal for teacher training.

Morado, M.F., Melo, A.E., & Jarman, A. (2021). Learning by making: A framework to revisit practices in a constructionist learning environment. *British Journal of Educational Technology, 52*(3), 1093–1115.

Resnick, M. (2006). Computer as paint brush: Technology, play, and the creative society. *Play= learning: How play motivates and enhances children's cognitive and social-emotional growth,* 192–208.

Resnick, M. (2017). *Lifelong kindergarten: Cultivating creativity through projects, passion, peers, and play.* MIT press.

Selwyn, N. (2009). *The digital native–myth and reality.* Paper presented at the Aslib proceedings.

Sun, L., Guo, Z., & Zhou, D. (2022). Developing K-12 students' programming ability: A systematic literature review. *Education and Information Technologies,* 27(5), 7059–7097.

Swan, K., Garrison, D.R., & Richardson, J. C. (2009). A constructivist approach to online learning: The community of inquiry framework *Information technology and constructivism in higher education: Progressive learning frameworks* (pp. 43–57): IGI global.

Swan, K., & Ice, P. (2010). The community of inquiry framework ten years later: Introduction to the special issue. *The Internet and Higher Education, 13*(1), 1–4. https://doi.org/10.1016/j.iheduc.2009.11.003

Tabarés, R., & Boni, A. (2023). Maker culture and its potential for STEM education. *International Journal of Technology and Design Education, 33*(1), 241–260. https://doi.org/10.1007/s10798-021-09725-y.

UNESCO. (2023). Artificial Intelligence in Education. Retrieved from https://www.unesco.org/en/digital-education/artificial-intelligence

Vartiainen, H., Tedre, M., Salonen, A., & Valtonen, T. (2020). Rematerialization of the virtual and its challenges for design and technology education. *Techne serien-Forskning i slöjdpedagogik och slöjdvetenskap, 27*(1), 52–69.

Webb, M., Davis, N., Bell, T., Katz, Y.J., Reynolds, N., Chambers, D.P., & Sysło, M.M. (2017). Computer science in K-12 school curricula of the 21st century: Why, what and when? *Education and Information Technologies, 22,* 445–468.

Congkak Quest
Enhancing STEM Skills through Game-Based Learning and Cultural Exchange

5

Zati Hakim Azizul Hasan,
Mas Sahidayana Mohktar, and
Siti Nursheena Mohd Zain

5.1 GAME-BASED LEARNING

Game-based learning (GBL) is an educational strategy that integrates game-play components into knowledge acquisition. It employs game design concepts to actively involve learners, stimulate their motivation, and improve their comprehension of educational material. GBL utilises interactive simulations, challenges, riddles, and narratives instead of traditional approaches such as lectures or textbooks to teach concepts and abilities (Tobias et al., 2014).

GBL possesses several fundamental attributes, such as commitment, motivation, adaptability, collaboration, simulation, and feedback (Figure 5.1).

DOI: 10.1201/9781003514626-5

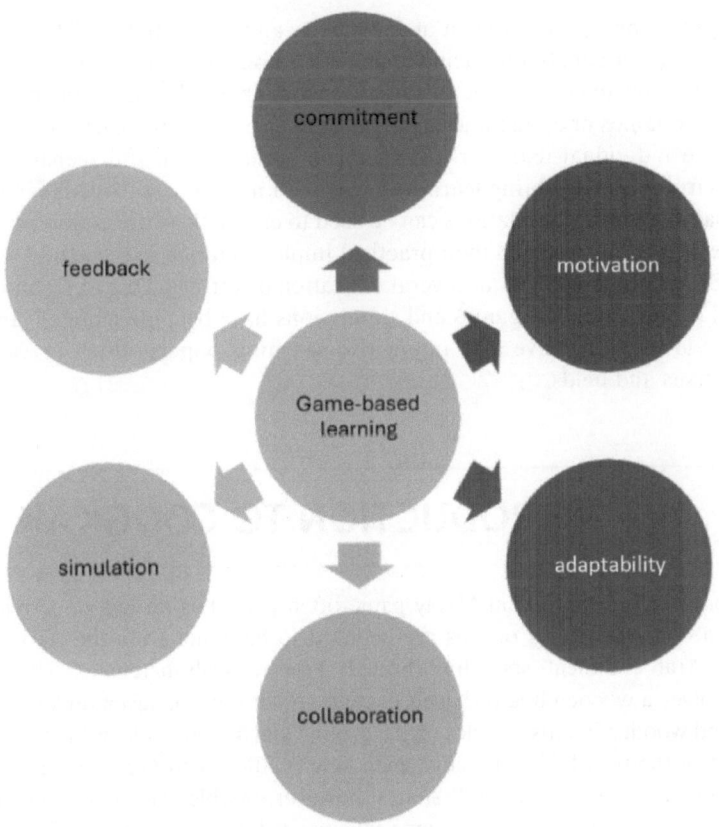

FIGURE 5.1   Fundamental attributes of GBL.

Games have an inherent feature that makes them highly engaging and able to maintain learners' focus for extended periods. GBL promotes active participation and enables deep learning by involving learners in interactive and dynamic settings. Games stimulate innate drive by introducing challenges, providing incentives, and cultivating a feeling of achievement. Players are often motivated to overcome problems and achieve goals in the game, resulting in increased engagement and perseverance in the learning process. Games offer immediate feedback to learners, allowing them to learn from their mistakes and make improvements in real time. This feedback loop enables continuous acquisition and mastery of skills. Many GBL platforms employ flexible algorithms to tailor the learning experience based on the learners' progress and performance. A personalised learning experience ensures that each student is given appropriate content in complexity and pace. Specific GBL experiences foster collaboration and teamwork among learners, nurturing social interaction and communication skills development. Participating in multiplayer games or collaborative projects can promote peer learning and cultivate individual teamwork. Games can simulate real-life scenarios and environments, enabling learners to use their academic knowledge in practical situations. Simulations can be used to establish a connection between theoretical notions and their practical implementation in the actual world. GBL can be employed in several educational settings. Due to technological progress, digital games and simulations have become more advanced, providing immersive and interactive learning opportunities in various courses and fields.

5.2 INTRODUCTION TO CONGKAK

Congkak is a traditional Malay game often played as an indoor activity or as a competition. Its origins are believed to be either from the African or the Arab continent, spreading through Asia by Arab merchants. The game involves a wooden board shaped like a boat with five or seven hollow perforated wooden boards or holes on the ground and a 'main hole or 'home' at the end of the board. Traditionally, each hole is filled with *buah* (game pieces) using rubber seeds or small stones; however, marbles are more commonly used today. This two-player game involves mental calculation and strategic moves to beat the opponent by bringing as many marbles home as possible to win the game. The name Congkak is believed to be derived from the Malay word *congak*, which means mental calculation. Congkak may have developed from Melaka in Southeast Asia as the state was a trading centre. Besides the Malays, the Congkak game is also famous among the descendants of Indians and Baba Malacca. Congkak is known as *Warri* or *Awari* in the Caribbean. In Indonesia, Congkak is known as *Congklak*, while in the Philippines it is *Sungka*.

Figure 5.2 shows the main tools of the Congkak game. The tools consist of the main element, the game board, a wooden board known as *papan Congkak* with a length of 80 cm–1 m and 18 cm–20 cm width. Inside the game board are holes called *kampung* or 'village.' These holes are arranged in two rows, each consisting of five or seven holes with a diameter of 6 cm. There are two big holes at the end of the rows with a diameter of 14 cm at both ends of the wooden board. The *guli* or seeds/marbles are needed to play the

RUMAH (HOUSE) KAMPUNG (VILLAGE)

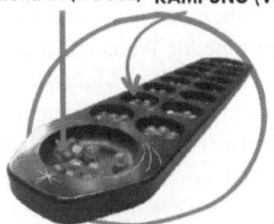

THE MAIN TOOLS

- **GAME BOARD** - A WOODEN BOARD KNOWN AS PAPAN CONGKAK 80CM TO 1M IN LENGTH AND 18CM TO 20CM IN WIDTH.

- **KAMPUNG (VILLAGE)** - HOLES ARRANGED IN 2 ROWS AND EACH ROW CONSISTS OF FIVE OR SEVEN HOLES WITH A DIAMETER OF 6CM.

- **RUMAH (HOUSE)** - BIG HOLES AT THE END OF THE ROWS WITH A DIAMETER OF 14CM ON THE LEFT AND RIGHT

- **GULI (SEEDS)** - THE NUMBER OF GULI PER HOLE DEPENDS ON THE TOTAL NUMBER OF HOUSES PER VILLAGE. HENCE, IF EACH VILLAGE CONSISTS OF 7 HOUSES, THEN THE GAME WILL START WITH 7 SEEDS PER HOUSE.

- ORIGINALLY USED CLAM SHELLS DURING THE EARLY DAYS.

FIGURE 5.2 The main components of a Congkak game.

Congkak. The number of seeds depends on the number of holes in the village. If there are seven holes in the village, the game will start with seven seeds per hole. Clamshells can also be used as seeds.

5.3 CONGKAK APPROACH TO STEM: A UNIVERSITI MALAYA STEM CENTRE CULTURAL EXCHANGE EXPERIENCE

Universiti Malaya (UM) STEM education's core aims are to educate and spark students' interest in STEM subjects. UM STEM education has been a long-standing endeavour to advocate for the importance of STEM knowledge and skills to meet the demands of STEM experts. The global pandemic COVID-19 has further ushered in the need for experts in the four pillars of STEM. Such demand has compelled STEM educators to reinvent and reimagine the approaches to engage students in STEM. One of the ways is through GBL, where traditional non-digital games are used as tools to impart knowledge and skills.

Four academics from UM STEM Centre, Malaysia, and one from BINUS University, Indonesia, have come together in the spring/summer to the UK under the Partnership in Employability (PIE) Programme for Women funded through the British Council's Catalyst Grant Application UK – Malaysia – ASEAN +3 Partnership Initiative: Catalyst Grants for Enhancing Graduate Employability 2022 and 2023. The visit is part of the British Council's global partnership initiative to nurture robotics research and science, technology, engineering, arts, mathematics, and health (STEAM-H) knowledge exchange between the PIE project partners of ASEAN and the UK.

The visit schedule included the academics engaging with underprivileged female students in Wales in a STEM cultural exchange activity at the National Museum Cardiff. A cohort of 28 female participants aged between 10 and 12 were present as part of the PIE's cross-culture and STEM women empowerment agenda. The academics of UM took the opportunity to deliver the notion of learning STEM through GBL in a collaborative effort with the PIE project leader from EUREKA Robotics Centre, Cardiff School of Technologies, Cardiff Metropolitan University. The UM STEM Centre provided a guidebook titled *STEM through Traditional Congkak Game*, along with plastic Congkak sets and marbles to groups of pupils.

The participants were acquainted with STEM education through the conventional Congkak game (Figure 5.3). Initially, Congkak may appear to be an unconventional instrument for introducing STEM education at UM. Nevertheless, it is an interactive tool designed to improve mathematical proficiency in addition and subtraction while requiring strategic thinking to win

FIGURE 5.3 Dr Zati Hakim Azizul Hasan, UM Program Leader for the PIE programme, explains the rules for playing Congkak.

the game by accumulating more seeds in one's *rumah* (home). Playing a game of Congkak can highlight numerous STEM traits, including critical-thinking and problem-solving abilities, which are crucial for addressing real-world issues.

Despite no prior knowledge of the game, the players were quickly acquainted with the rules with minimal guidance from the facilitators. Both teachers and students were engaged in the competition to get the most seeds in their respective Congkak houses. They were observed having fun; notably, 26 of the 28 players could complete several simple mathematical scenarios to test their understanding of the game. Therefore, the impact of a simple traditional game was not only meaningful as part of a cultural exchange activity, but, more importantly, deemed a suitable GBL tool for the upscaling STEM skillsets for all age groups, particularly children. At the end of the sessions, the Congkak boards were given to both schools, so they could continue to utilise this game as part of the teaching and learning experience in the classroom.

5.3.1 Revival of the Congkak and the Future

Sadly, the value of Congkak has perhaps been forgotten among the locals due to the advent of digital gaming. This and the declining number of students taking STEM subjects over the years motivated the PIE UM team to revive this traditional game after the positive outcome seen with the children from Wales with Congkak.

A roadshow in Malaysia was conducted to empower underprivileged children, particularly girls in STEM, via the introduction of Congkak GBL and robotics. The participation was overwhelming, with 100 underprivileged children and 60 girls from Program Perumahan Rakyat (PPR) Kg. Baru Hicom taking part in the first event, followed by a second event involving 101 male and 103 female students from SMKA Sultan Azlan Shah, Seri Iskandar, Perak, Malaysia. Surprisingly, there was interest from students up to 17 years old, indicating that Congkak cuts across all ages (Figure 5.4).

The roadshow also showcased robotics and drones through three modules: obstacle avoidance navigation with Lego Mindstorm, Yolo and OpenPose Vision Module, and Patrol with AlphaBot and drone display.

5.3.1.1 Obstacle avoidance navigation with Lego Mindstorm

Using a prebuilt Lego Mindstorm, this station showcased autonomous navigation through sonar and bumper sensors (Figure 5.5). The seamless avoidance

FIGURE 5.4 Students at SMKA Sultan Azlan Shah, Seri Iskandar, Perak, Malaysia, focused on their Congkak game.

FIGURE 5.5 Testing obstacle avoidance in outdoor environments.

of collisions sparks immediate curiosity among children and draws the children's attention to understand the way it works behind the robot's intelligence, igniting a passion for STEM education.

5.3.1.2 Yolo and OpenPose vision module showcase

This station displays real-time object detection and pose estimation results featuring robots equipped with YOLO (You Only Look Once) and OpenPose vision modules (Figure 5.6). This setup allows kids to interact with the robots, place objects and observe how the robots perceive and understand their environment through advanced vision technology.

5.3.1.3 Patrol with Alphabot and drone display

This station presents cutting-edge robotics technologies to children, offering a glimpse into the future of real-world robotics applications. Alphabot, which can be controlled via phone for surveillance and the drone display, showcases how technology safeguards our surroundings. Together, the Alphabot and drone inspire a sense of wonder and curiosity among children and motivate them to explore these fields further (Figure 5.7).

FIGURE 5.6   Testing pose estimation accuracy with YOLO detector.

FIGURE 5.7 Underprivileged children at PPR Shah Alam experiencing robotics coding and drone applications.

5.3.2 The science 'S' of playing Congkak

Engaging in a game of Congkak can foster the development of fundamental scientific process skills. The Congkak player must rely on their visual and tactile capabilities to gather information about the seeds. At the start of the game, the Congkak players must separate and arrange the seeds according to the established rules. Players collect numerical data on the quantity of seeds. In order to explain an event or observation, they must make logical initial inferences, which may be accurate or inaccurate. In order to anticipate the next move, individuals must rely on previous observations and experiences, as well as take into account the number of seeds present in each Congkak hole.

The scientific process skills acquired from the Congkak GBL can be used to learn the school's science subjects, especially in the laboratory sessions. GBL facilitates students' early engagement with the complex concepts inherent in the lab structure, enabling them to grasp their complexities (Huebra et al., 2020).

5.3.3 The technology 'T' of manufacturing Congkak

Congkak is typically constructed from timber. Currently, plastic Congkak is also accessible. When creating a Congkak board, computer numerical control (CNC) machines can automate wood cutting, carving, and moulding according to digital designs. Woodworkers can utilise Computer-Aided Design (CAD) and Computer-Aided Manufacturing (CAM) software to generate intricate digital Congkak designs and simulate the manufacturing process before commencing production. This technology optimises the design process, enhances precision, and enables customisation. Although not widely used in woodworking compared to other industries, 3D printing technology can produce tailor-made jigs, moulds, and prototypes for Congkak manufacturing. This technology enables quick development and experimentation with complex forms and patterns. Timber drying and preservation technologies can expedite the drying process, minimise flaws, and augment the longevity of wood-based Congkak products.

5.3.4 The engineering 'E' of Congkak GBL educational content design

Developing the Congkak GBL guidebook, 'STEM Through Traditional Congkak Game,' involved several key stages in the engineering process. First, we conducted a needs assessment and identification of learning objectives. We identified the target audience, their learning needs, and their specific educational objectives. Then, we developed the Congkak game concept and storyline to align with learning objectives, ensuring they facilitate skill acquisition and understanding of concepts (Figure 5.8). We gathered feedback from educators, subject matter experts, and potential learners to improve gameplay and learning outcomes. Monitoring, evaluation, and feedback are ongoing to identify areas for improvement and inform future iterations of the game.

5.3.5 The mathematics 'M' skills of playing Congkak

Playing Congkak can enhance players' mathematical proficiency across various mathematical ideas. The Congkak game facilitates the provision of context and motivation for learning. The game will enhance the player's fundamental arithmetic abilities, including addition, subtraction, multiplication, and division. The players must possess a comprehension of statistics, the capacity to interpret data, and a grasp of probability. In addition, they will need to employ logical reasoning in order to formulate problem-solving techniques. In addition, the players will have to answer mathematical questions in the provided worksheets (Figure 5.9) after playing the Congkak games.

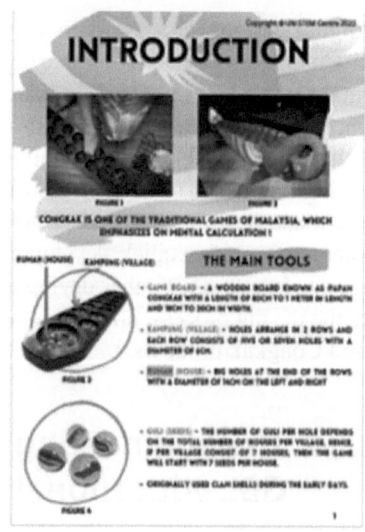

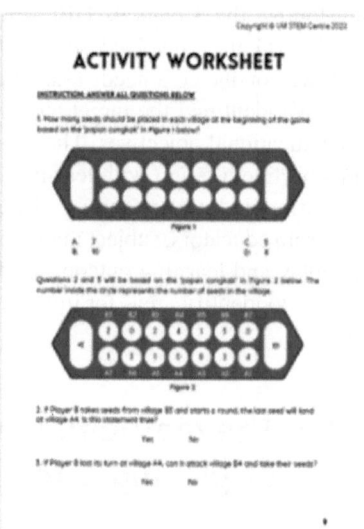

FIGURE 5.8   Some of the content from the STEM Through Traditional Congkak Game guidebook.

There are eight mathematical questions provided in the Congkak guidebook. The students were given time to calculate and discuss the answers. Most of the students answer 7 out of 8 questions correctly. The last question required the students to think critically; consequently, most did not get the answer correct on the first attempt.

FIGURE 5.9   Players answer the mathematical questions after playing the traditional Congkak game.

5.4 REIMAGINING GBL 'CONGKAK AND COBOT': A HUMAN–ROBOT INTERACTION STEM INNOVATION

Collaborative robots, or cobots, are purposefully created to securely engage with humans in a shared environment. Cobots equipped with dual arms enhance adaptability and agility while executing activities necessitating synchronised manipulation or concurrent operations. This design enables them to imitate human-like movements and carry out a broader spectrum of jobs. Dual-arm cobots are outfitted with user-friendly interfaces and communication technologies that enable effortless contact with human operators. These features encompass touchscreen interfaces, speech recognition, gesture control, and augmented reality interfaces, which enhance the ease of human interaction and control over the cobots (Peshkin & Colgate, 1999).

Cobots can potentially revolutionise traditional games like Congkak by enhancing various aspects of the game experience. We foresee humans engaging in a game of Congkak using dual-arm collaborative robots in a GBL setting. In order to emerge victorious in the game, a human player must employ their mathematical acumen in domains such as addition and subtraction while also employing strategic thinking. Cobots can serve as interactive GBL-Congkak tutors, teaching beginners or younger players the rules and tips for strategic moves. As players improve, the cobot can increase the difficulty level and speed of play to keep the game engaging and enjoyable.

There are several methods available to train cobots in the skill of playing Congkak. Reinforcement Learning (RL) methods can facilitate the robot's learning process through trial and error by engaging with the environment. Some examples of RL techniques include Q-Learning, Deep Q-Networks (DQN), and Proximal Policy Optimization (PPO). Supervised learning can be employed to train cobots in accurately predicting the appropriate actions to be taken in various scenarios in the Congkak game. This method necessitates a substantial dataset of human gameplay instances for training. Neural networks, decision trees, and support vector machines are prevalent methods used in supervised learning. Imitation learning is an alternative approach where the robot learns by observing human gameplay and replicating the shown behaviour. Methods such as behavioural cloning or inverse reinforcement learning are utilised to instruct the robot in playing games by mimicking human participants. Cobots can also be trained using evolutionary algorithms. This technique emulates the process of natural selection to maximise the robot's behaviour across numerous generations. Genetic algorithms, genetic programming, and evolutionary techniques can be employed to develop game-playing strategies. Bayesian inference enables the cobots to revise their beliefs regarding the game state by considering observable evidence and prior knowledge. This methodology is advantageous for making probabilistic determinations in games with uncertainty or insufficient information (El Zaatari et al., 2019).

Soft robotic grippers can be engineered with pliable and adaptable materials to imitate human fingers, enabling them to extract seeds or marbles from the Congkak board. Soft robotic grippers possess flexibility and compliance, enabling them to conform to various forms and sizes of objects. This adaptability allows them to efficiently grip things, including unusual shapes or delicate surfaces. Soft robotic grippers can manipulate various things, encompassing rigid and flexible items, without necessitating intricate programming or modifications. Their adaptability makes them highly suitable for activities that require the manipulation of various objects in uncertain surroundings. Soft robotic grippers apply lower levels of force on objects than their rigid counterparts, hence minimising the likelihood of causing harm to fragile things. They are well-suited to jobs that involve delicate manipulation, such as extracting seeds or marbles from the interior of the Congkak board. Figures 5.10–5.12 display many instances of pliable robotic grippers.

This soft robotic gripper results from a platform technology developed by Harvard researchers to create soft robots with embedded sensors that can sense inputs as diverse as movement, pressure, touch, and temperature. (Credit: Ryan L. Truby/Harvard University)

Inspired by biology, the TU/e soft robotics lab focused on addressing fundamental challenges such as mobility, control, dexterity and haptic perception through soft bodies made of hyperelastic materials.

New 3D-printed soft robotic gripper functions without electronics. The device is printed all in one go and can pick and release objects.

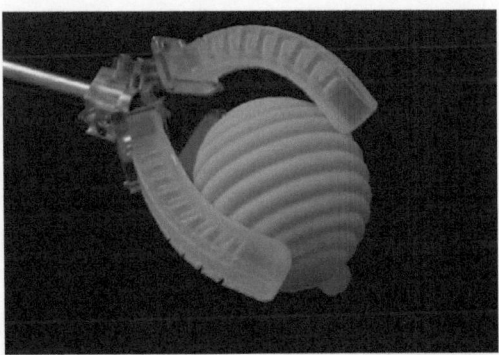

FIGURE 5.10   Soft robotic gripper is the result of a platform technology developed by Harvard researchers, adopted from Crowe (2018).

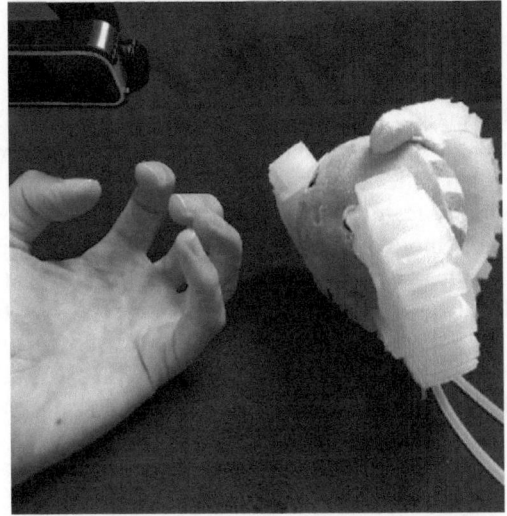

FIGURE 5.11   Biology-inspired soft robotics gripper adopted from Soft Robotics (n.d.).

During the PIE programme, a knowledge-sharing session was conducted with Professor Perla Maiolino, the Soft Robotics Research Group Lead at Oxford Robotics Institute. She shared her experience developing a modular soft robotic arm with omnidirectional bending capabilities using 3D printing (Ouyang et al., 2022).

Having Congkak in digital platforms and integrating with cobots can expand the GBL-STEM approach with Congkak to other cultures. Cobots

FIGURE 5.12   3D-printed soft robotic gripper adopted from Sakharkar (2023).

can potentially analyse gameplay data, highlighting players' STEM weaknesses and strengths. The feedback can help players improve their STEM skills and develop better Congkak strategies over time. Cobots can incorporate accessibility features such as voice commands or gesture controls, making the game more accessible to players with disabilities or mobility impairments. Inclusivity can enable players from different backgrounds, cultures, and STEM skills to interact with their peers in a fun and engaging manner. By leveraging the capabilities of cobots, Congkak can be transformed into a more immersive, accessible, and engaging game that appeals to a broader audience and preserves its cultural significance in the digital age.

REFERENCES

Crowe, S. (2018, August 7). *3D Printing Soft Robotics with Embedded Sensors*. The Robot Report. https://www.therobotreport.com/97810-2/

El Zaatari, S., Marei, M., Li, W., & Usman, Z. (2019). Cobot programming for collaborative industrial tasks: An overview. *Robotics and Autonomous Systems, 116*, 162–180.

Huebra, M., Ibarretxe, J., Okariz, A., Sarasola, A., & Zubimendi, J. L. (2020). Game-based learning of scientific skills. *In EDULEARN20 proceedings* (pp. 2052–2057). IATED.

Ouyang, W., He, L., Albini, A., & Maiolino, P. (2022, April). A modular soft robotic arm with embedded tactile sensors for proprioception. In *2022 IEEE 5th International Conference on Soft Robotics (RoboSoft)* (pp. 919–924). IEEE.

Peshkin, M., & Colgate, J. E. (1999). Cobots. *Industrial Robot: An International Journal*, *26*(5), 335–341.

Sakharkar, A. (2023, September 20). *New 3D-printed soft robotic gripper functions without electronics*. Tech Explorist. https://www.techexplorist.com/3d-printed-soft-robotic-gripper-functions-without-electronics/65655/

Soft Robotics. (n.d.). https://www.tue.nl/en/research/research-groups/dynamics-and-control/soft-robotics

Tobias, S., Fletcher, J.D., & Wind, A.P. (2014). Game-based learning. *Handbook of research on educational communications and technology*. Springer.

Implication, Challenges, and Moving Forward

6

Esyin Chew and Anwar P.P. Abdul Majeed

6.1 COMPENDIUM: THE PIE PROGRAMME

The book provides a comprehensive overview of the efforts to promote gender equality and diversity in STEM fields through education and outreach programmes across different regions, including Wales, Malaysia, and Indonesia. It emphasises the importance of addressing the severe underrepresentation of women in science, technology, engineering, and mathematics (STEM) globally and presents various initiatives and strategies to bridge this gap primarily via the Partnership for Innovation in Employability (PIE) programme, which is proudly supported by the British Council, UK.

Chapter 1 sets the stage by introducing the background and significance of gender equality in STEM. It highlights the persistent gender gap in STEM fields, with women accounting for only around 30% of the world's researchers and even fewer in disciplines like engineering and maths. The chapter then introduces the PIE programme, which aims to foster women's engagement with STEM through collaborations between universities and industry

DOI: 10.1201/9781003514626-6

partners in the UK, Malaysia, and Indonesia. The programme focuses on enhancing the employability of women in STEM, developing interdisciplinary courses, and instilling essential skills for women pursuing STEM careers.

Chapter 2 delves into the EUREKA STEM Robotics and Artificial Intelligence (AI) educational programme in Wales, which targets children and young people from underprivileged communities. The chapter examines the transformative journey of the EUREKA Robotics Centre at Cardiff Metropolitan University (CMET) and its initiatives to promote STEM education and inclusivity. Through collaborations with global stakeholders, the centre has played a significant role in enhancing robotics and AI literacy, developing skills, and empowering individuals from diverse backgrounds to pursue STEM careers. The chapter also highlights the centre's evolution into the dynamic STEAM (science, technology, engineering, arts, and mathematics) Hub and its impact on fostering innovation and driving meaningful societal changes in Wales and beyond.

Chapter 3 shifts the focus to Malaysia, where Universiti Teknologi MARA (UiTM), a PIE partner, has been actively promoting STEM education among children and teenagers through various outreach programmes. Following the Quadruple Helix Engagement model, UiTM collaborates with the government, industry, and community to bring about positive changes in knowledge, attitudes, skills, and aspirations. The chapter highlights several initiatives, such as the Maker Talent for Digital Innovation (MTDI) and the STEM4Fun programme, which provide training, workshops, and competitions to foster interest in STEM fields, particularly among underprivileged and rural students. These programmes aim to establish a sustainable ecosystem for hands-on learning experiences and instill a culture of innovation.

Chapter 4 explores the challenges, best practices, and future direction of integrating digital-making skills into STEM education in Pahang, Malaysia. It is worth noting that Pahang is on the east coast of Malaysia where STEM literacy and resources are not as exposed nor accessible as those in the central part of Malaysia. Through a Community of Inquiry (CoI) workshop spearheaded by UMPSA STEM Lab, Universiti Malaysia Pahang Al-Sultan Abdullah (UMPSA), another PIE partner, engage with educators. The chapter examines the importance of digital-making skills in fostering creativity, innovation, and problem-solving abilities among students. It also addresses the challenges faced by educators in teaching these skills, such as limited resources, diverse student skill levels, and the need to balance traditional and technology-driven teaching methods. The findings reveal strategies for overcoming these challenges, including adopting individualised teaching approaches, incorporating hands-on activities and real-world applications, and fostering collaborative learning environments. The chapter emphasises the importance of professional development opportunities for educators and collaborations with external stakeholders to enhance digital making education. It also envisions the future of digital making education, involving the integration of emerging technologies like AI, virtual reality (VR), and augmented reality (AR) and a focus on practical applications of digital skills beyond the classroom.

Chapter 5 introduces an innovative approach to promoting STEM education through game-based learning (GBL) using the traditional Malay game Congkak. The effectiveness of Congkak as a GBL tool for developing critical-thinking and problem-solving skills among underprivileged female school children was demonstrated through this chapter. The chapter explores the various aspects of STEM that can be learned through playing Congkak, including scientific process skills, the role of technology in manufacturing Congkak boards, the engineering process involved in designing the GBL guidebook, and the mathematical skills that players can develop. The chapter concludes by envisioning a future where Congkak can be integrated with collaborative robots (cobots) and AI decision-making to create an immersive and accessible digital platform that appeals to a wider audience while preserving the game's cultural significance.

6.2 REFLECTION ON THE OVERALL IMPLICATIONS OF THE PIE PROGRAMME

Following the sentiment of quoting Sidelil et al.'s (2023) from the beginning of this book that, *'Being in science and at the same time being a woman is difficult': Academic women's experiences of gender inequalities in STEM*, we cohesively share four impact stories in Wales and Malaysia throughout the book from the diverse perspectives of women STEM stories and how we make little impact on the world around us. These efforts would not happen without the great financial and catalyst support from the British Council and the Welsh Government. Innovating between eight PIE partners from three countries (UK, Malaysia, Indonesia) from 2022 until present, the PIE programme has produced one Collaborative Agreement among all six partners and has now evolved to be a self-sustained Global PIE programme, based in Wales and serving the world. After years of efforts of Partnership for Innovation in Employability (PIE) programme for women in STEM, we achieved beyond what we proposed at the beginning, including:

1. The development and signing of 6 Memorandum of Agreements (MoAs) with participating Malaysia universities to seal the collaborative relationship until December 2028, meeting the UK

Research Excellent Framework (REF 2029) for the international impact case study.

2. The successful PIE I student and staff mobility between UK and Malaysia for 19 people within two years. The PIE I programme doubled the planned UK–Malaysia mobility within the project period: initially planned for at least 8 pax for mobilities, but in reality we delivered 19 mobilities with CMET's Alan Turing Institute co-cash matched funding. The PIE II programme strengthened the PIE I project capacity, enhanced existing partnerships by sharing interdisciplinary (healthcare care and agriculture) knowledge, and exchanged best practices among eight partners with global mobility: 46 senior management, academics and students in 2023–2024 between UK and Malaysia for PIE II. In total 65 mobilities and placements between UK and Malaysia have been facilitated by the PIE programme.

3. Informed and joint curriculum with four guest lectures were delivered for MSc in Robotics & AI based on industry feedback and the PIE programme; we also officially created two new postgraduates in co-supervision for Masters in Computer Science (Research).

4. Within two years of the PIE programme, we collaboratively applied for research and development bids to the sum of £1,691,504.34 with a 23% success rate and attracted a total grant value of £386,504.34 to date.

5. The PIE programme received widespread attention, featuring in more than 30 media outlets across national and international platforms. This coverage extended to various institutional social media channels and was prominently displayed on the project's official website.[1]

6. A summative PIE event at Cardiff School of Technologies to showcase the findings and outcomes of the PIE programme through strand presentation and interactive display stands. The summative event brought together representation from all higher and further education institutions as well as a range of external organisations. The event involved strand presentations from the educational institutions and the industry partners involved, a facilitated PIE panel, and an exhibition space for networking and to demonstrate the success of the programme. We successfully delivered the summative 2-day PIE event at Cardiff School of Technologies, CMET in 2023: (i) Day 1 with the honours of the presences of SEC GENs of both Malaysia Minister of Higher Education (YBHG. Dato' Seri Abdul Razak Bin Jaafar) and Welsh Government Higher Education department (Sinead Gallagher), Head of Global Partnership British Council Wales (Brenda Giles), Vice-Chancellors, Pro-Vice Chancellors, Deans and key PIE programme members from Malaysia and Wales universities for MoAs signing; (ii) Day 2

with the WiDEN and PIE Panel Discussion for Employability and Leadership: Where are the Women? Mapping the job gap in AI, a policy briefing led (online) by Professor Kate Royse, Director of Hartree Centre part of the UK Science and Technology facilities Council (STFC) National laboratories and Professor Kirstine Dale, Principal Fellow/Co-Director for Joint Centre for Excellence in Environmental Intelligence and with Malaysian PIE academics and students, and EUREKA researchers in the audience. PIE for Women blended keynotes speech by Professor Perla Maiolino, the Lead of the Soft Robotics Research Group, the only female lead at Oxford Robotics Centre (2024) and Dr Barry Bentley from EUREKA Robotics Centre (2024).

7. In addition, PIE II STEAM-H 101 and 102 produced joint efforts of STEM-robotics courses with certification provision and seminars for traditionally underrepresented groups (see Figure 6.1 of the new model) to engage with AI for STEAM-H (science, technology, engineering, arts, mathematics, and healthcare), through widening discourse and public understanding of the technical benefits and risks of their AI applications in solving the real-world problems, innovative pedagogy, competency-based curriculum and digitalisation for B40 students and non-STEM specialist

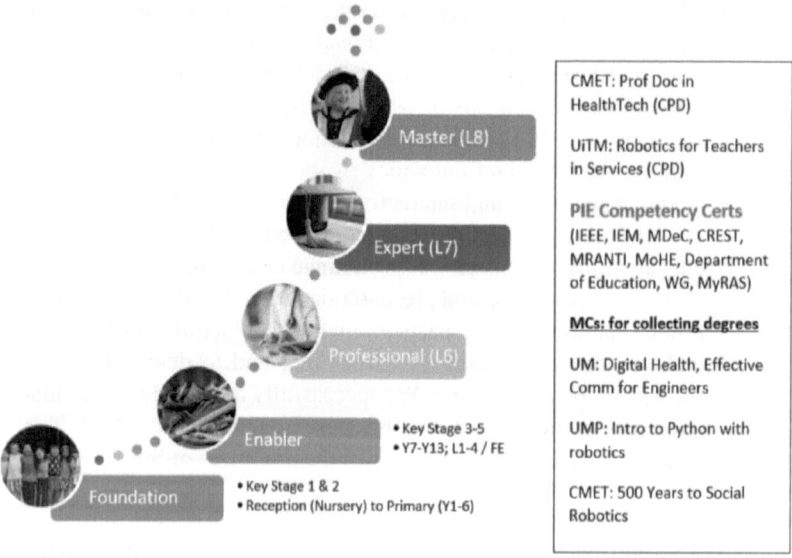

FIGURE 6.1   PIE – joint efforts of STEM-robotics courses with certification provision and seminars.

audiences. We have further developed dual-PhD, the first in CMET and new framework with formal procedures approved by the institutional Graduate School and Quality Education Development department. Supervisors from PIE universities (UiTM, Universiti Malaya (UM) and University of Science Islam Malaysia (USIM)) are interviewed and recruited for co-supervising the first batch of dual-PhD candidates between Wales and Malaysia from the next academic years. This is high-quality joint research that has strong links with industry and relevant professions, and serves as an extensive opportunity for personal and academic career development. This is interdisciplinary PIE research and transnational education in robotics in healthcare and agriculture (dual-PhD/ professional doctoral programme), through supporting global mobility for staff and students with the following achievements:

- Within two years of the PIE programme, we collaboratively applied for research and development bids to the sum of £1,658,272.34 with a 23% success rate and attracted a total grant value of £378,272.34 to date.
- Total value of £869,848.33 from 14 joint bids for PIE II with successful projects worth £340,356.80 – 48.6% success rate out of the known results.
- Successful delivery of 32 robotics workshops, impacting over 6,000 people.
- Over 10 joint papers produced among the PIE partners.
- Two EUREKA Nexus/Satellite Lab setup in Malaysia with six Tokku Zones (ALTY, 2022–2023; Hu & Chew, 2024).

The above joint work leads to the short-term impact of a wider participation of the international placement through collaborative supervision. This had an immediate impact for the participating women in STEM, as well as for the hosts. It allows women in STEM to be able to access cutting-edge facilities and equipment at the host institutes in both the UK and Malaysia. The PIE programme also provided robot and AI work experience in the field at care homes and hospitals in both the UK and Malaysia that will enhance the participants' employability in related fields. Employability skills, high-level training, and work-based learning is the direct upskilling for the participating women. The mid-term impact in the next year would be the joint Women in STEM workshop deliveries in both countries that will inspire young girls to study in STEM subjects with joint certifications for internationalisation. The shared knowledge and employability consultancy services will be a strong provision of the PIE programme. A new dual-PhD programme and joint supervision in postgraduates' curriculum in robotics and AI are designed to further support skills development and international STEM employability in areas of critical economic and industrial importance, and academic collaboration between the participants and partner institutes.

6.3 DISCUSSION OF CHALLENGES ENCOUNTERED AND RECOMMENDATIONS FOR ADDRESSING CHALLENGES

With an honest reflection, we wish to share the below challenges of the PIE programme for the reference of any international partnership or interested practitioners:

1. **Cultural Diversity:** different institutions have different scholarly cultures and academic requirements, i.e. co-authorship and the quantitative vs qualitative approach of research papers, which can sometimes cause minor conflicts during meetings. CMET need to manage various expectations and emphasised the originality and academic integrity culture, bridging disagreement and obtaining mutual understanding and consent. Cultural sensitivity between the UK and Malaysia academics and students shall be managed and briefed before any mobility. Be passionate for the country you serve and are based in, and pre-learn the local culture, be professionally respectful and inclusive from the bottom of the heart to receive international visitors or to visit a foreign country during the mobility.

2. **Regulatory and legal necessity**: sometimes we may feel that it took too long for some MoAs to be reviewed by respective legal teams in partnering universities. Patience, trust, and frequent communication are the key values we uphold in the PIE programme and beyond.

3. **Disparity of equality in higher education institutions**: a dual-PhD mobility grant was offered to UM, the best research university in Malaysia, but was pulled at last minute and offered to extended PIE II member USIM, as the project team learnt that UM has a policy that only sends PhD scholars to the top 50 (25) QS ranked higher education institutions and CMET does not meet the criteria. This is a debate about the 'Elite educational system ethos' vs 'Equal Education,' and about the meaning and value of equality of

educational opportunity based on one QS ranking. The fundamental questions could be, 'Is equality of opportunity achieved when everyone with similar talent gets the same results regardless, they attend QS-top-ranked universities or middle-ranked? When those with the same natural talent potential get the same opportunities in dual programme in the UK and Malaysia?' (Alexander, 1985; Anderson, 1999)

4. **High tuition fees in the UK universities for developing countries:** all partners expressed that the tuition fees of both postgraduates and PhD students are too high and not affordable for Malaysians of average incomes and underprivileged B40 pupils. They value dual-PhD that can help in reducing two years of living costs and tuition fees; however, even the one year of tuition fees at CMET is not affordable. The PIE Project Lead is attempting to apply from the senior management of CMET to explore the options of offering home student fees, as well as scholarships or externally funded PhDs. The Project Lead also attempted to apply for British Council scholarship and status of Doctoral Training Centre, but both were unsuccessful.

5. **Communication:** some young academics from the UK or Malaysia universities maybe be complacent or overconfident in their interpretation of 'quality research.' Mutual respect is highlighted in all meetings and private communication by the key leaders to ensure that the expectations are managed with regards to research outcome. As the Project Lead, CMET attempted to bridge the communication gap and educate participants on the expectations of research quality from the UK-REF's perspective of not merely indexed journal outputs, but rather (1) people and culture that matters; (2) the actual contribution to knowledge and understanding of the joint research work; and (3) real-life engagement and impactful case study – all three are the key REF (2019) influenced PIE ethos in the global partnership.

Further recommendations including top-down and peer-influenced are the essence of our successful PIE programme. Bottom-up can emerge, but may take longer for overseas partnerships in our case. The support from the senior management team including vice-chancellor or president of a university can accelerate the sealing of a partnership. In the PIE programme, we highlight the 'Women-Human approach,' indicating motherly and sisterly communication, i.e. 1-2-1 communication and coaching seems to be the most effective. People-oriented research culture, regardless of gender and race (women-only groups and meetings sometimes could be undermined or over mindful).

6.4 TUNNELLING AHEAD: PIE VISION FOR THE FUTURE AND AVENUES FOR FURTHER COLLABORATION

Yorke (2004) suggests that employability is 'a set of achievements – skills, understandings and personal attributes – that makes graduates more likely to gain employment and be successful in their chosen occupations, which benefits themselves, the workforce, the community and the economy.' We believe that PIE has advance the PIE members' employability by widening participation and rich international exposure. Although the funding of PIE I and II has ceased, PIE has nevertheless received new grants to expand further to Pakistan, Indonesia, and China, as well as continuing the joint healthcare robotics research and STEAM intervention for underprivileged communities in both Wales and Malaysia. The long-term impact in the coming years will be more collaborative research development among the partners with joint grant applications to respective countries' grants, building a long-term internationalisation mechanism to increase commitment and strategic partnerships to address shared priorities, personnel development, and shared expertise in STEM between universities in the UK, Malaysia, Indonesia, Pakistan, and China. This will act as a catalyst to inspire senior management in respective universities to offer women in STEM scholarships for the development of the next generation, with beneficiaries being:

- Female undergraduates, postgraduates, and PhD students in Malaysia and Wales.
- Women in engineering, computer science, and academia.
- Public and school awareness raising for women in STEM.
- Malaysia public universities to promote Women in STEM initiatives.
- Welsh Government and UK Parliament due to the incoming talent and socioeconomics benefits. Programme impacts, including written evidence, can be an international reference on the benefits of the successful PIE model.

UN (2023) highlights that the contribution of women is incalculable, and we strongly echo that PIE for women in STEM and STEAM-H are indeed immeasurable. In Chapter 1, we reflected that through cross-cultural education with sociological imagination and history (Wright, 1959), the gender equality and diversity are evolved in the past and present. However, Wright further differentiates 'reason (critical and reflexive thought – closer to freedom)' and 'rationality (associated with organisation and efficiency – a lack of reason).' There may be some rationale behind why men leadership in Malaysia universities is common and men's applications for academic promotion are better articulated than women's applications. Alternatively, the reasons why always women senior leaders are easier to throw out or be disrupted by family commitment.

In the eyes of Wright (1959), those women academics who do not exercise reason and passively accept their social position is referred to as the 'Cheerful Robot' in which the individual is alienated from the unequal society: this is a pressing concern, as the Cheerful Robot is the 'antithesis' of democratic society; it's the 'ultimate problem' of gender equality and as a threat to society's values. On the other hand, Europe has the highest gender parity of all regions at 76.3%, with one-third of countries in the region ranking in the top 20, and 20 out of 36 countries have at least 75% parity. Iceland, Norway, and Finland are the best-performing countries (World Economic Forum, 2023). Young et al. (2021) reveal that there are extensive disparities between women and men in skills, status, pay, seniority, industry, job, attrition, and educational background and therefore call for effective policy responses if society is to reap the benefits of technological advances.

CMET has one of the highest numbers of female university vice-chancellors in the past seven years. The PIE programme, based in CMET, is dedicated to the safe evolution of social and educational humanoid robotics. Its research laboratories include the STEAM Hub, an intelligent robot-maker lab for 3D design and robotic parts printing. Research activities are also carried out within the centre's Autonomous Robotics Lab for the development of remotely operated systems. Its HRaaS Hub, which is evolving service robotics for healthcare and hospitality, is funded by the British Council, Welsh Government, Engineering and Physical Sciences Research Council (EPSRC), and the Alan Turing Institute (EUREKA Robotics Centre, 2024). Together with Global PIE partners, we hope to better tackle global and national challenges. Although many countries with higher levels of gender equality have not successfully managed to dismantle career barriers for women in science (Going Global, 2023), we persistently aim to empower women and lower social-economic communities, as suggested in the UN sustainable development goals, the pathways to gender equality as integral scientific research. We are realising this goal with internationally joint disruptive educational and healthcare robotics research by building resilient infrastructure, and promoting inclusive and sustainable innovation for women academics and those in healthcare sector.

Paradoxically, The World Economic Forum's 2023 Global Gender Gap Report predicts that it will take a whopping 131 years at the current rate of progress to achieve full gender parity. The STEM labour market continues to be an area of particular challenge, where women remain significantly underrepresented in the STEM workforce, especially at leadership levels. How can the global higher education sector accelerate closing the gap? (Going Global, 2023, p. 12). We would affirm that change can begin with knowledge exchange, and efforts are being made at all levels of multi-institutional project in multilateral countries, as demonstrated, to encourage girls and young women to excel in STEM employment, regardless in academia or industry (PIE, 2024). The historical PIE events have helped our shape of today and the lives of every living person. As such, PIE learning history can help us view our lives within the context of others in the future PIE, based on past experiences. We hope this book provides you with a better basis for understanding gender equality and how the PIE actions can be valuable in impacting your new little world order through the lens of women with social imagination.

NOTE

1   https://pie4stemwomen.wixsite.com/pie4stem

REFERENCES

Alexander, L.A. (1985). 'Fair equality of opportunity: John Rawls' (Best) forgotten principle,' *Philosophy Research Archives*, 11: 197–208. doi:10.5840/pra19851111

ALTY Hospital (2022–2023). Robot EUREKA received the Sultan (State King) in Malaysia. https://twitter.com/eurekarobot/status/1545355526866755586?s=20&t=6lFiuuWG3z2-DEUdPBA3CQ; The First Tokku Zone in Malaysia by ALTY Hospital: https://www.facebook.com/altyortho/photos/a.119072853568490/462786712530434/?type=3; https://www.businesstoday.com.my/2023/06/07/alty-orthopaedic-hospital-cardiff-university-to-establish-malaysias-first-tokku-zone-in-a-single-specialty-hospital; https://www.linkedin.com/posts/chewesyin_alty-orthopaedic-hospital-cardiff-university-activity-7075391142451654656-KyeW?utm_source=share&utm_medium=member_desktop http://mrem.bernama.com/viewsm.php?idm=46291

Anderson, E.S. (1999). 'What is the point of equality?. *Ethics*, *109*(2), 287–337. doi: 10.1086/233897.

EUREKA Robotics Centre. (2024). EUREKA Robotics Centre, Cardiff Metropolitan University. www.cardiffmet.ac.uk/eureka

Going Global. (2023). Conference Session Details. British Council https://www.britishcouncil.org/sites/default/files/session_list_v0.4_final.pdf

Hu, S., & Chew, E. (2024). EUREKA's Tokku Zones published by UK Parliament: https://committees.parliament.uk/writtenevidence/128382/pdf/

Oxford Robotics Centre. (2024). Oxford Robotics Centre, Oxford University People List https://ori.ox.ac.uk/people-list

PIE. (2024). PIE I and PIE II programme, funded by British Council. https://pie4stemwomen.wixsite.com/pie4stem/clients

Sidelil, L.T., Spark, C., & Cuthbert, D. (2023). Being in science and at the same time being a woman is difficult': Academic women's experiences of gender inequalities in STEM academia in Ethiopia. *Women's Studies International Forum*, 98. https://doi.org/10.1016/j.wsif.2023.102717

UN, United Nations (2023). UNESCO in Action for Gender Equality, UNESCO 2023 https://www.unesco.org/en/gender-equality

World Economic Forum. (2023). Global gender gap report 2023: Insight report. https://www3.weforum.org/docs/WEF_GGGR_2023.pdf

Wright, M. (1959). *The sociological imagination* (reprinted 2000). Oxford University.

Yorke, M. (2004). *Employability in higher education: What it is – What it is not*. The Higher Education Academy/ESECT.

Young, E., Wajcman, J., & Sprejer, L. (2021). Where are the women? Mapping the gender job gap in AI. chrome-extension://efaidnbmnnnibpcajpcglclefindmkaj/ https://www.turing.ac.uk/sites/default/files/2021-03/where-are-the-women_public-policy_full-report.pdf

Index

Note: *Italicized* page references refer to the figures and **bold** references refer to the tables.

3D modelling, 49, 62

A

Aberystwyth Robotics Club, 19
Absidee Group Sdn Bhd, 43
Age group, *57*
AI, *see* Artificial Intelligence (AI)
Alan Turing Institution, 22
Alan Turing Public Engagement Grant
    award, 23, 26
Alphabot, 75, 77
Aquaculture, 10
AR, *see* Augmented reality (AR)
AR Distributor, 43
Arduino microcontrollers, 27, 34, 35
Artificial Intelligence (AI), 4, 16–27, 62, 87;
    see also Alphabot; Augmented
    reality (AR); EUREKA; Machine
    learning techniques; Virtual
    reality (VR)
    aim of, 22
    careers in STEM fields, inspired by, 26
    decision-making, 18, 88
    education, 17–20, 22, 25, 27, 87
    historical background on, 18–20
    for STEAM-H, 90
Asas Sains Komputer (ASK), 47, 50,
    60–61, 67
Augmented reality (AR), 62, 81, 88
Australian Academy of Science, 3
Autonomous Robotics Lab, 17, 95
Azoulay, Audrey, 7, 12

B

Bentley, B., 90
BINUS University, 8, 74
BioAI-BioEngineering Lab, 17
Bowes-Lyon, Elizabeth, 2
Bray Leino Limited, 24

British Council, 11, 22, 24, 74, 86, 88, 89,
    95; *see also* Partnership for
    Innovation in Employability (PIE)
    programme

C

CAD, *see* Computer-Aided Design (CAD)
CAM, *see* Computer-Aided Manufacturing
    (CAM)
Cam, Helen Maud, 2
Cardiff Metropolitan University (CMET), 8,
    9, 11, 16–18, 89, 91–93, 95
Cardiff School of Technologies, 9, 17, 74, 89
Catholic Church dogma, 1–2
CDYDIP, *see* CelcomDigi Young Digital
    Innovators Programme
    (CDYDIP)
Celcom Axiata Berhad, *see* CelcomDigi
    Berhad
CelcomDigi Berhad, 38–39, *39*, 42
CelcomDigi Young Digital Innovators
    Programme (CDYDIP), 38–42,
    41, *42*
Chew, E., 19–21
Child Friendly Cardiff, 24
Chumbaka Sdn Bhd, 35
CMET, *see* Cardiff Metropolitan University
    (CMET)
CNC, *see* Computer numerical control
    (CNC)
Cobot, 81–84
Coding, 46–49, 52, 59, 61, 64, 65, *78*
CoI, *see* Community of Inquiry practice
    (CoI)
Commonwealth Games, 24
Communication, 48, 72, 93
Community of Inquiry practice (CoI), 52–53,
    53, 87
Computer-Aided Design (CAD), 79
Computer-Aided Manufacturing (CAM), 79

Computer numerical control (CNC), 47, 78–79
Computer science, 50–52
Congkak/Congklak, 70–84, *73*, *74*, *76*, *80*, *81*
  Alphabot, 75, 77
  approach to STEM, 73–75
  and cobot, 81–84
  components of, *73*
  concept of, 79
  drone display, 75, 77, *78*
  engineering, 79
  game-based learning (GBL), 70–72, *71*, 73–75, 79, 81–84, 88
  guidebook, 74, 79, 80, *80*, 88
  introduction to, 72–73
  Lego Mindstorm, 75–77
  mathematics, 79–80, *81*
  OpenPose Vision Module, 75, 77
  revival and future of, 75–77
  science, 78
  technology, 78–79
  Universiti Malaya (UM), 73–75
  You Only Look Once (YOLO), 77
Continuing professional development (CPD), 19, 48, 64, 67
COVID-19 pandemic, 4, 5, 40–41, 73
CPD, *see* Continuing professional development (CPD)
Culture, 1, 16, 22, 25, 33, 74, 87, 92, 93
  academic integrity, 92
  diversity, 92
  of innovation, 25, 33
  people-oriented, 16, 93
  technological, 22
Curriculum design, 9

D

Deep Q-Networks (DQN), 81
Dewey, J., 46
Digital innovation, 32–34
Digital Maker Hubs (DMH), *34*, 34–35
Digital making skillsets, 46–67, *58*, *62*
  application of, 63
  Asas Sains Komputer (ASK), 60–61
  Community of Inquiry practice (CoI), 52–53
  computer science, 50–52
  education development, 46–47
  hypotheses, 54–55, 65–66
  physical computing, 49–50

programming, 42, 47, 49–50, 59, 61, 62
  RBT school, 58–60
  research methodology, 53–55
  Sains Komputer (SK), 47, 61
  teaching
    challenges, *58*, 58–61
    practices, 62–65
    programming, 49–50
    technology design, 50–52
    UMP STEM Lab, *47*, 47–49
DMH, *see* Digital Maker Hubs (DMH)
DQN, *see* Deep Q-Networks (DQN)
DreamCatcher Sdn Bhd, 35
DroBotics, 42–44, *43*
Drone display, 75, 77, *78*

E

EDS, *see* IEEE Electron Device Society (EDS)
Education, 3, 8, 17, 19, 22–25, 33, 44, 46–67
  AI, 17–20, 22, 25, 27, 87
  development, 46–67
  higher, 1–5, 11, 18, 19, 22, 27, 31, 34, 92–93
  in Pahang, 55, 65–66
Embedding technology, 9
Employability toolkit, objects of, 9; *see also* Partnership for Innovation in Employability (PIE) programme
Employable learner, 9
Employment, 25, 26, 27, 94
  PIE programme, 9
  in STEM fields, 5, 96
  women, 26
Empowering the Future: Digital Making Skill Sets in STEM Education in Pahang Malaysia, 53
Engineering and Physical Sciences Research Council (EPSRC), 95
Ethical-Ubiquitous-Robotics driving Economy Knowledge Accelerator for Wales (EUREKA), *see* EUREKA
EUREKA, 16–27, 87, 90, 91
  careers in STEM fields, 26
  early policy influences of, 20–22
  historical background on, 18–20
  impact on STEM education, 22–24
  introduction, 16–18
  transformation, 24–27

EUREKA Robotics Centre, 8, 9, 11, 16–27, 74, 87, 90; *see also* Partnership for Innovation in Employability (PIE) programme
EZ robots, 20

F

First Tech Challenge, 18
Fourth Industrial Revolution, 21

G

Game-based learning (GBL), 70–72, *71*, 73–75, 79, 81–84, 88; *see also* Congkak
GBL, *see* Game-based learning (GBL)
Gender
    disparities, 25
    equalities, 1–7, 8, 24, 86, *see also* UNESCO
    gap, 3
    inequality, 2
    roles, 4
    violence based on, 6
Gender Equality Admissions Program, 3
Global Gender Gap Report, 95–96
Global Grants Scheme, 25
Griffiths, Annie Hughes, 2

H

Harvard researchers, 82, *83*
Harvard University, 2
Hasan, Zati Hakim Azizul, *74*
Healthcare/hospitality, 8–10, 18, 21, 26, 91, 94
    applications, 10
    professionals, 23, 25
    robotics, 9, 17, 95
    women in, 8
HEFCW, *see* Higher Education Funding Council for Wales (HEFCW)
HESA, *see* Higher Education Statistics Agency (HESA)
Higher Education Funding Council for Wales (HEFCW), 19
Higher education institutions, 92–93
Higher Education Statistics Agency (HESA), 4
HRaaS Hub, 17, 95
Huang, Jesen, 18

Humanoid robots/robotics, 17, 19–20, 24, 27, 95
    history of, 20
    social/educational, 17, 95

I

ICT, *see* Information and communication technology (ICT)
IEEE Electron Device Society (EDS), 43–44
Information and communication technology (ICT) Career of Choice (ICTCoC) programme, 34–35
Innovation competitions, 36–37, *37*
Intelligent Robot (IR), 17, 95
Internet of Things (IoT), 39, 62
IoT, *see* Internet of Things (IoT)
IR, *see* Intelligent Robot (IR)

J

Jobs, Steve, 18
Johor Corporation, 37

K

Kampung, defined, 72
Khan, Shadan, 24

L

Leader/leadership, 4, 10, 21, 25–27, 90, 95, 96
Learning, 4, 19, 40–41, 64; *see also* Teacher/teaching
Lego Mindstorm, 75–77
Let's go to Mummie's Lab (LGTML), 11
LGTML, *see* Let's go to Mummie's Lab (LGTML)

M

Machine learning techniques, 17; *see also* Artificial Intelligence (AI)
Maiolino, Perla, 83, 90
Maker bootcamps, 37–38, *38*
Maker Talent for Digital Innovation (MTDI), *33*, 33–39, *39*, 42, 87
    CelcomDigi Young Digital Innovators Programme (CDYDIP), 38–42
    defined, 33
    DroBotics, 42, *43*

innovation competitions, 36–37, *37*
maker bootcamps, 37–38
MDEC ICT Career of Choice, 34–35
programmes, 33–42
Young Innovate Programme, 35–36, *36*
Malaysia Digital Economy Corporation
(MDEC), 33–35
Malaysian Education Development Plan, 31
Malaysian Ministry of Education, 42
Malaysian public schools programmes, **51**
Massachusetts Institute of Technology
(MIT), 2
McLuhan, M., 20
MDEC, *see* Malaysia Digital Economy
Corporation (MDEC)
Memorandum of Agreements (MoAs), 88,
89, 92
Microcontroller, 34–35, 38, 39, *39*, 41
Microsoft, 18
Ministry of Communications and Digital, 33
MIT, *see* Massachusetts Institute of
Technology (MIT)
MoAs, *see* Memorandum of Agreements
(MoAs)
MPhil/PhD programmes, 9, 11, 91–92
MSC, *see* Multimedia Super Corridor (MSC)
MTDI, *see* Maker Talent for Digital
Innovation (MTDI)
MTDI Space, defined, 34
Multimedia Super Corridor (MSC), 33

N

Nao robots, 20
National Museum Cardiff, 20, 74
NGOs, *see* Non-governmental organisations
(NGOs)
Non-governmental organisations (NGOs),
39, 43

O

Obstacle avoidance, *76*
Oxford University, 2

P

Pahang, 55, *56*, 65–67
Papan Congkak, defined, 72
Partnership for Innovation in Employability
(PIE) programme, 8–12, 24, 74,
74, 75, 83, 86–96, *90*

aims of, 8–9, 10
challenges of, 92–93
compendium, 86–88
fundamental rationale of, 9
implications of, 88–92
PIE I, 8–9, 22, 27, 89, 94
PIE II, 8, 10, 22, 27, 89, 90–92, 94
vision and future of, 94–96
Payne-Gaposchkin, Cecilia, 2
PEDi, *see* Pusat Ekonomi Digital (PEDi)
PHP, 41
Physical computing, 49–50
PIE programme, *see* Partnership for
Innovation in Employability (PIE)
programme
Piscopia, Elena Lucrezia Cornaro, 1–2
PLL, *see* Polygenic-weLLbeing (PLL)
Polygenic-weLLbeing (PLL), 25
PPO, *see* Proximal Policy Optimization
(PPO)
PPR, *see* Program Perumahan
Rakyat (PPR)
Programme/programming, 42, 49–50, 59,
61, 62; *see also* Partnership for
Innovation in Employability (PIE)
programme; RBT programmes
CDYDIP, 38–42, *41*, *42*
and digital making skillsets, 47, 49, 50
DroBotics, 42
genetic, 82
ICTCoC, 34–35
languages, 41, *see also* PHP; Python;
SCRATCH; SQL
Malaysian public schools
programmes, **51**
MPhil/PhD, 9, 11, 91–92
skill, 47
STEM4Fun, 43–44, 87
STEM Ambassador, 19
teaching, 49–50
Young Innovate, 35–36, *36*
Program Perumahan Rakyat (PPR), 75, *78*
Proximal Policy Optimization (PPO), 81
Pusat Ekonomi Digital (PEDi), 42
Python, 41

Q

Q-Learning, 81
Quadruple Helix Engagement model, 31–32,
32, 43, 44, 87
Questions formulation, *62*, *64*

R

RBT programmes, 50, 58–60, 67
  primary school, 58–59
  secondary school, 59–60
REF, *see* Research Excellent Framework
  (REF)
Regulatory and legal necessity, 92
Reinforcement Learning (RL), 81–82
Rekabentuk Teknologi (RT), 46
Research Excellent Framework (REF),
  89, 93
Rights Fest Cardiff, 24
RL, *see* Reinforcement Learning (RL)
Robots/robotics
  and Artificial Intelligence (AI), 8–9,
    17–27, 87, 91
  coding, 78
  degrees/courses/workshops, 20, 21, 90
  educational, 25
  humanoid, *see* Humanoid robots/robotics
  Nao, 20
  soft gripper, 82, 83, 84
Robot Xiaolongbao, 25
Rogers, Annie, 2
RT, *see* Rekabentuk Teknologi (RT)

S

Sains Komputer (SK), 47, 61
Science, technology, engineering,
    agriculture, mathematics, and
    healthcare (STEAM-H), 8, 23,
    74, 90, 94
Science, technology, engineering, and
    mathematics (STEM), 1, 32–33,
    47, 86; *see also* Education; MTDI;
    Quadruple Helix Engagement
    model; Women
  Congkak, 70–84
  context setting, 10–12
  continuing professional development
    (CPD), 19
  decadal plan for Women in, 3
  education, 46–67, *see also* Digital
    making skillsets
    analysis of hypotheses, 65–66
    findings, 58–65
    research methodology, 53–57
    revolution of digital making for,
      46–53
  employment in, 5–6

EUREKA, 16–27, 87, 90, 91
  gender equalities and diversity in, 1–7
  gender gap in, 3
  introduction, 8–10
  leadership in, 4, 25–27, 96
  Learning, 4, 19
  Maker Talent for Digital Innovation
    (MTDI), 33, 33–39, 39, 42, 87
  overview, 1–7
  Partnership for Innovation in
    Employability (PIE) programme,
    8–12, 22, 24, 27, 74, 74, 75, 83,
    86–96
  robotics courses, 90
  role of, 53
  Universiti Teknologi MARA (UiTM), 8,
    11, 31–44, 87, 91
  women in, 1–12, 87–89, 91, 94
  workshops, 8, 19, 23, 26
Science, technology, engineering, arts,
    mathematics (STEAM), 17–19,
    24–27, 87, 95
SCRATCH, 41
Sidelil, L.T., 88
SK, *see* Sains Komputer (SK)
Skills, 50, 53, 54, 58–62, 62, 63, 67, 79,
    81, 87
  coding, 46–49, 52, 59, 61, 64,
    65, 78
  communication, 48, 72
  entrepreneurial, 9
  logical thinking, 49
  mathematical, 79–80, 88
  presentation, 41
  problem-solving, 37, 47–49, 62–65, 67,
    75, 79, 87, 88
  programming, 47
  reasoning, 18, 49, 79
  STEM, 48, 75, 84
  teaching, 87
  technical/soft, 38
SME, *see* Subject matter expertise
    (SME)
Societal changes, 20, 24–27, 87
Soft robotic gripper, 82,
    83, 84
SQL, 41
SSDU Innovations Sdn Bhd, 37
Stanford, Jane, 2
STEAM, *see* Science, technology,
    engineering, arts, mathematics
    (STEAM)

STEAM-H, *see* Science, technology, engineering, agriculture, mathematics, and healthcare (STEAM-H)
STEM, *see* Science, technology, engineering, and mathematics (STEM)
STEM4Fun programme, 43–44, 87
STEM Ambassador Programme, 19
STEM Learning, defined, 19
STEM-robotics courses, *90*
STEM through Traditional Congkak Game, 74
Subject matter expertise (SME), 55, *56–57*
Subject taught in class, *56*
Sungka, *see* Congkak
Swallow, Ellen, 2

T

Teacher/teaching, 25
    CDYDIP webinar series, 40
    challenges, *58*, 58–61
    Community of Inquiry practice (CoI), 52–53
    continuing professional development (CPD), 19
    digital techniques, 40
    experience, *57*
    methods, 54, 58–63, 66, 67, 87
    plans, 58
    practices, *62*, 62–65
    programming, 49–50
    Sains Komputer (SK), 61
    skills, 87
Technical and Vocational Education and Training (TVET), 11
Technology design, 50–52
Tenaga Nasional, 37
Tuition fees, 93
TVET, *see* Technical and Vocational Education and Training (TVET)

U

UiTM, *see* Universiti Teknologi MARA (UiTM)
UK Department for Education, 18
UM, *see* Universiti Malaya (UM)
UMPSA, *see* Universiti Malaysia Pahang Al-Sultan Abdullah (UMPSA)
UMP STEM Lab, 47–49, *48*, 87
UNESCO Global Priority Gender Equality Framework, *6, 7*

United Nations Educational, Scientific and Cultural Organization (UNESCO), 6, *6,* 7, *7*, 8, 10–12
Universiti Malaya (UM), 8, 73–75, *74*, 91, 92
Universiti Malaysia Pahang Al-Sultan Abdullah (UMPSA), 8, 11, 47–48, 87
Universiti Teknologi MARA (UiTM), 8, 11, 31–44, 87, 91; *see also* Maker Talent for Digital Innovation (MTDI)
    Quadruple Helix Engagement, 31–32, *32*, 43, 44, 87
    STEM4Fun, 43–44
University of Chile, 3
University of Stanford, 2
University of York, 8, 9
US Census Bureau, 5

V

Virtual reality (VR), 62, 63, 66, 88
VR, *see* Virtual reality (VR)

W

Warri /Awari, *see* Congkak
Welsh Government, 20, 22, 23, 88, 89, 94, 95
Welsh League of Nations Union (WLNU), 2
Widening support, 9
Williams, Ivy, 2
WLNU, *see* Welsh League of Nations Union (WLNU)
Women
    in Africa, 3
    in Asia, 4
    in Australia, 3
    decadal plan for, 3
    employment, 26
    healthcare, 8
    higher education to, 11
    history, 1–2, 10–11
    Latin America, 3
    leadership, 4, 10, 26, 27
    in Middle East, 4
    in STEM, 1–12, 87–89, 91, 94
World Economic Forum, 2–3, 95–96
World War II, 2
Wright, M., 94–95
W-STEM project, 3

Y

YOLO, *see* You Only Look Once (YOLO)
Yorke, M., 94
York Robotics Lab, 9

Young, E., 95
Young Innovate Programme, 35–36, *36*
You Only Look Once (YOLO), 75, 77, *77*
YouTube, 40